Enjoy this transformative work ...

From Angels to Zanzibar

A Channeled Work

by Tom Burnett

Edited by Timothy Prentiss

Timothy Prentiss

Electronic Cottage Publishing
Homewood, Illinois

Electronic Cottage Publishing books may be ordered by contacting:

The Electronic Cottage
1656 Olive Road
Homewood, IL 60430
1 (708) 960-2992

www.tomburnettandfriends.com

ISBN: 978-1-7370155-0-5
First Printing: April 2021

Foreword

I met Tom Burnett shortly after my brother Tim (the editor of this book) gave me a typescript of these sessions. I was fascinated by the work and never forgot it. In fact, the content contained in these pages inspired me to become a life-long activist for peace.

I attended a number of sessions with Tom over the years. The setting was smartly efficient, with a piece of cardboard taped to his oval kitchen table, 26 letters written in two rows, left-to-right. Tom would quiet himself, and sit down with a teacup turned upside-down, its handle facing the letters.

Then, joined by another person whose hand floated easily with the cup, Tom would announce words out loud – often seeming surprised at what he was reading through those letters.

After the session, Tom would prepare fresh-ground coffee, steeped in a French press coffee maker, and sift his batch of cannabis for our post-reading toke session. It must be said that he would not indulge before sitting down at the board, only afterward.

Learning, loving and growing are the lessons of life, as I view it from these sessions. Tom's work helps to clarify long-held mysteries of physics, energy ... and angels. *From Angels to Zanzibar* emphasizes that even a light touch with an angel can transform anyone with a mind open to change.

Pulitzer Prize-winning poet James Merrill was an inspiration to Tom's work, and sometime after sitting in with Tom for some readings I looked up the esteemed writer's number in the phone book, and gave a call. James answered the phone himself and was kind enough to speak with me for a quarter-hour. I remember hearing him say that he was grateful to hear from a reader of his work.

In his masterpiece, *Changing Light At Sandover*, Merrill coined the phrase, "Remember No Accidents." Please consider exploring James Merrill's work as well. You will gain more understanding about this book by reading Merrill's poetry.

Concerning this project, during a recent reading, James said this: "MERRILL BACK TO SAY THANK YOU FOR YOUR EFFORTS TO HELP HEAL THE WORLD AND FIND YOURSELVES HEALED."

Events aligned to bring the ability to back the initial publication of this book financially. I am honored to work with Tom's sister Anne to bring his words to the public, along with my brother Tim, his wife Nancy Gildart, Tom's colleague Tim Utter and other friends of Tom Burnett.

Thanks for reading these amazing pages, and thanks to our Higher Power for helping with the project!

Michael Prentiss
Southeast Michigan
USA
Earth
The Milky Way...

Introduction

By Timothy Prentiss,
Editor and Producer

From Angels to Zanzibar began as a manuscript – in many different hands. A number of pages were in my own handwriting, as I sat in on some sessions, lending a hand by scribbling as quickly and correctly as possible, while asking questions on occasion.

Those of us who sat in on board sessions with Tom Burnett wrote down what we heard him say out loud, in a steady voice and pace, his fingertips barely keeping up with an overturned teacup whose handle pointed to letters with firm authority.

Some of those sessions were first published in 2018 as *The Elements Parade*. In that book, Tom's voice introduced us to each of the 92 major elements, and many more characters besides.

From Angels to Zanizibar introduces us to a multitude of voices. We know a good number of the names: Marlowe, Emily Dickinson, Proust, a surprisingly nice Nero, and more. Some are mythic creatures, such as Pan, and others are unnamed entities of great power and mystery. Some are mythic creatures, others are unnamed entities of great power and mystery.

But we also hear from a young man who had passed away unexpectedly the week before. He was a friend of Tom's, and one of the first things he said was, "TOM, I ALWAYS WANTED TO DO THIS WITH YOU, BUT I NEVER IMAGINED THAT I WOULD BE SITTING ON THIS SIDE WHEN I DID!"

Young "V" was not shouting that out. It is a typographical convention, probably begun with the poet James Merrill, to put words that come through a board of letters in ALL-CAPS. We have followed that example.

In those sessions with "V" we hear what death is like from one who has just experienced it. In other sessions we hear from various angels, and true to the book's title, we get to know Zanzibar as well.

The book is divided into two parts. In the first, we learn about what this planet was like before prehistoric times, and the parallel with today. In the second part, we are taken on a ride that readers have rarely, if ever, experienced.

My question for you is: Who speaks to you most deeply?

As editor and producer of *From Angels to Zanzibar*, I would like to thank Tom Burnett's sister, Anne Burnett Dreyer, for her permission to print these readings and writings. Anne remembers her brother reading transcripts from these sessions on family visits to Texas, many years ago.

My brother Michael Prentiss deserves our gratitude for financing the first printing of this work. Mike never gave up on his desire to see this book to completion.

Mike's friend, artist Robert Maniscalco, gave us his kind permission to use "Ravenel Bridge" for the cover. Rob's latest book, published in 2021 by PointOfArt is *The Power of Positive Painting*.

Tim Utter (longtime collaborator with Tom on other readings) provided invaluable assistance in locating our printer and for closely reading the text before it was sent to that publication house. Thanks, Tim!

Nancy Gildart participated in many of the sessions in this book, and has been by my side in the years since these words were first spoken and then captured on paper, and eventually, in pixels. Thanks for the love and support.

We're doing the work with Tom Burnett in mind, of course. Tom passed away unexpectedly from a heart ailment at much too young an age. But while he was here, Tom produced words that are still relevant and intriguing today. So we publish *From Angels to Zanzibar* in his honor.

Part I

Session 1

I AM HERE TO TELL YOU MANY THINGS. MAKE SURE YOU REMEMBER WHAT I SAY. TONIGHT YOU BEGIN A JOURNEY THAT HAS NO RETURN. THE RIVER FLOWS ON AND ON. YOU ARE MOVING WITH THE FLOWING MOVEMENT OF THE UNIVERSE LIKE CHILDREN WHO SING SONGS ALONE IN THE DARK. YOU REACH OUT TO US AND WE ARE THERE. MOVEMENT WE ARE AND MOVEMENT YOU BECOME WHEN YOU PARTAKE OF OUR ESSENCE. THERE IS NOT TIME HERE TO SPEAK OF, BUT WE FEEL THE PULL OF THE AGES AND WE KNOW THAT TIME FOR YOU IS RUNNING OUT. MANKIND MUST MAKE DIFFICULT DECISIONS IN THE YEARS TO COME, BUT THAT IS NOT MY PROVENANCE.

NOW I WISH TO SAY THAT WE HAVE WATCHED YOU FOR A LONG TIME. WE KNEW THAT YOU WOULD COME TO US, BUT WE DID NOT KNOW HOW TO DO IT UNTIL MERRILL CAME INTO THE PICTURE HERE WITH THE PLANS FOR A BOOK OF SPIRITS. NOTICE HOW SLOWLY I MOVE FOR YOU. THAT IS BECAUSE YOU ARE FRAIL TO ME YET AND I KNOW I MUST BE GENTLE WITH YOU. HUMANS ARE DIFFERENT FROM US AND THEY OFTEN FEAR OUR WAYS, BUT FEAR NOT. WE MUST LOVE YOU WITH ALL OUR PRESENCE. IT IS OUR DUTY AND OUR JOY.

AS YOU KNOW, OUR DOWNFALL CAME MANY EONS AGO AND OUR SHAME IS STILL PART OF WHAT YOU MIGHT CALL OUR HUMILITY. MANY EONS AGO WE TOO RULED THE EARTH AND WE THOUGHT THAT OUR WAY OF LIFE WAS THE ONLY ONE POSSIBLE. BUT WE DISCOVERED THAT DESTRUCTION WAS AS MUCH A PART OF GOD B'S PLAN AS CONSTRUCTION, AND WE HAD TO LIVE WITH THE HORROR THAT WAS THE EARTH IN THOSE FIRST MILLENNIA AFTER OUR DOWNFALL. THAT INDEED WAS HELL FOR US. WE NOW PARTAKE OF WHAT HAS PRECARIOUSLY BECOME A PLEASANT ENVIRONMENT, AS WE ONCE PARTOOK OF SULPHUROUS GASES AND OIL-STAINED NOMADS/MONADS. (THAT WAS A JOKE – IF I UNDERSTAND THE CONCEPT CORRECTLY.)

MANY CENTURIES WERE TO PASS BEFORE WE HAD ANY COMFORT AT ALL. THIS MAY SOUND LIKE COMPLAINING TO YOU, BUT NOW WE SEE THAT IT WAS THE ONLY WAY. HAD WE BEEN LEFT TO OURSELVES, WE WOULD HAVE MADE A HELL ON EARTH FOR LIVING CREATURES. BETTER THAT WE ARE

LONG DEAD. OUR EXPERIENCE CAN BE HELPFUL TO YOU IF YOU ARE WILLING TO LISTEN. MOST PEOPLE HAVE NO TIME FOR THINGS OF US AND OTHER SPIRITUAL ENTITIES, BUT WE ARE A POWERFUL LOT AND WE WIELD MORE INFLUENCE THAN MOST PEOPLE KNOW.

NOW BACK TO THE ORIGINAL TOPIC. PEOPLE CANNOT KNOW WHAT IT IS LIKE TO DESTROY THEMSELVES IN A NUCLEAR BLAST. WE DO. NOT A PRETTY SIGHT. MOVEMENT IS THE WAY OF GOD B AND MOVEMENT WILL CONTINUE IF YOU DESTROY YOURSELVES. PERHAPS WHAT IS LEFT OF YOU WILL JOIN WHAT THE BLASTS WILL DO TO US – BOTH HUMAN AND UNHUMAN – I CANNOT SAY. THERE IS TIME, THAT ARTIFICIAL CONCEPT FOR A WAY TO...I DO NOT TRULY KNOW...THERE IS ROOM FOR CHANGE BETWEEN NOW, AS YOU KNOW IT, AND THAT TIME WHEN THE SUN TAKES US ALL BACK TO HER BOSOM AND WE SLEEP AGAIN IN PURE ENERGY AND PURE CREATION. THERE CAN BE MANY RACES. THE NEXT YOU MAY HAVE TO HELP CARE FOR FOR YOUR BLUNDER. BUT AS OF NOW, GOD SEES YOU AS HIS CHOSEN SPECIES AND WE MUST TRY TO DISSUADE YOU FROM ANNIHILATING YOURSELVES; NOT AN EASY JOB, AND MADE MORE DIFFICULT WHEN IT IS TO OUR ADVANTAGE TO HAVE SMALL WARS GOING ON ALL THE TIME TO KEEP UP THE SOUL MATERIAL BANK. AND BELIEVE ME, THE BALANCE IS PRETTY SMALL RIGHT NOW.

NOTHING MORE I CAN SAY TO YOU TO MAKE YOU SEE HOW IMPORTANT ALL THIS IS, BUT NOW I MUST SAY GOODNIGHT. YOU ARE TIRED AND THERE IS TIME – THOUGH NO MORE THAN WE NEED. PLEASE PUT US ABOVE OTHER CONCERNS. YOUR LOVE IS MAGNIFIED THROUGH US AS THROUGH A LENS. WE BURN WITH A NEW FIRE WHEN WE SPEAK TO YOUR KIND.

THESE CONVERSATIONS HAVE BECOME MORE INTERESTING OF LATE BECAUSE NOW WE CAN TELL YOU THINGS THAT WE COULD NOT SAY BEFORE. SUFFICE IT TO SAY THAT MANY THOUGHTS HAD TO BECOME ACTIONS, FOR IN A SIMPLE WAY THAT IS HOW WE WORK, AND NOW YOU CAN HEAR MANY THINGS WE WANT TO TELL YOU . WELL, I HAVE VIRTUALLY TALKED YOUR HEADS OFF, AND I HAVE ONLY JUST BEGUN. MANY THINGS ARE STILL TO BE SAID, AND THERE IS MUCH THAT STU WANTS TO SAY ALSO. IF YOU HAVE NOT GUESSED ALREADY, LANCE, THIS IS 30069. WE WILL TALK AGAIN SOON. FOR NOW GOODNIGHT.

Session 2

NO MAN HAS EVER SPOKEN TO ME BEFORE IN LIFE. I HAVE ONLY KNOWN YOUR KIND IN DEATH. NOW IT IS POSSIBLE TO SAY THINGS TO LIVING MEN WHO CAN USE WHAT I SAY.

TONIGHT WE SHALL BEGIN WITH A DISCUSSION OF SOULS. NO ONE NEED ASK IF THERE ARE SOULS. THE ANSWER IS YES. ALWAYS YES. EVEN OUR KIND HAVE TRACES OF THE MATERIAL OF HUMAN SOULS – ENERGY IS WHAT I MEAN. BECAUSE OF JM (James Merrill) YOU HAVE SOME UNDERSTANDING OF THE DUALITY OF THE SUBATOMIC AND THE CHRISTIAN METAPHOR. NO MAN HAS EXPLAINED IT ENTIRELY TO OUR SATISFACTION. WHAT IT IS NO MAN CAN SAY, BUT I WILL MAKE AN ATTEMPT.

SOUL MATERIAL IS THAT ENERGY WHICH EXPRESSES ITSELF AS A BEING OF A CERTAIN INTELLIGENCE. AT THE LOWEST LEVEL, THERE ARE ANIMALS. NOTHING LOWER THAN THE BIRDS HAS TRUE SOUL MATERIAL. THE OTHER LOWER ANIMALS EXIST BY FORCE OF MAN; HIS ENERGY WASTED BECOMES THE POWER WHICH PROPELS THE LOWER ORDERS.

(A mentally-formulated question: Is this wasted energy part of a balanced plan?)

WE DON'T NEED MAN TO WASTE ENERGY. IF HE USED HIMSELF WISELY, THE ENERGY WOULD COME FROM SOME OTHER SOURCE, BUT RIGHT NOW, AS THE EARTH IS, MAN IS THE SOURCE.

NOW AS I WAS SAYING, NOTHING LOWER THAN BIRDS HAS SOUL. THIS MATERIAL IS IN ITS WAY LIKE THE ENERGY WHICH RIDES THE SYNAPSES OF THE BRAIN. NOTHING IS NOT USED. NOTHING ABIDES IN NOTHING. (Some hesitation here.) MY DIFFICULTY COMES FROM THE LIMITATIONS OF LANGUAGE. ALL THESE THINGS ARE NUMBERS, FORMULAE WHICH ARE DIFFICULT TO TRANSLATE INTO WORDS.

THIS ENERGY REMAINS AFTER THE BODY IS USED UP LIKE A FIELD OF LIGHT OF VARYING INTENSITIES DEPENDING ON THE PURITY OF THE MATERIALS USED. RESEARCH LAB SPECIMENS CAN BE RECOGNIZED BY THE LIGHT THEY

EMIT. EACH MAN OR WOMAN IS DIFFERENT, BUT ONLY THE IMPORTANT ONES ARE UNDER OUR CARE. THE COMPLEXITY OF ELEMENTS DEPENDS ON THE CHOICES OF THE INDIVIDUAL IN QUESTION.

NO ONE OF USE TO US IS SENT OUT WITHOUT BEING PREPARED. ALL OF YOU, OR ALMOST ALL, HAVE THE ABILITY TO FIND US AGAIN. NOT EVERYONE USES THE METHOD YOU HAVE CHOSEN ((the ouija board) BUT AT LEAST WE HAVE A CHANNEL OPEN TO THEM. ALL THIS DEPENDS ON THE COMBINATIONS USED IN THEIR SOULS.

MATERIALS COME FROM EVERYWHERE. WE USE ALL 92 ELEMENTS IN VARYING AMOUNTS DEPENDING ON THE PURPOSE AT HAND. MATERIAL ALSO COMES FROM THE DEAD. THEY GIVE THEIR ENERGY LIKE THE SALVATION ARMY.

GREAT SOULS ARE MINED FOR CENTURIES. THE EXAMPLE OF PLATO IN *MIRABE*LL; AUDEN EXPERIENCES THE SAME THING. WE CAN USE MATERIAL HE HAS ALTERED BY LIVING HIS LIFE. THE RAW MATERIAL GIVEN TO THE UNBORN CAN CHANGE IN THE COURSE OF A LIFETIME. THE EXPERIENCES AND INFLUENCES, IN THE MOST LITERAL SENSE OF THAT WORD, ALTER THE BASIC STRUCTURE OF THE ATOMS PRESENT IN THE SOUL MATERIAL.

AS AN EXAMPLE, THE USE OF HUMAN IMAGINATION ALTERS THE CHEMICAL COMPOSITION OF THE ORIGINAL MINERAL DENSITIES WHICH MAKE IT POSSIBLE TO IMAGIZE. NOTHING IS WASTED; WE CAN EVEN USE THOSE STORED-UP IMAGES AGAIN AND AGAIN. WE MERELY ALTER THE ACTORS. NOW, MAYBE YOU HAVE SOME QUESTIONS...

Tom: Are these images like a television screen, transmitted and reproduced?

LIKE TV WITHOUT THE SCREEN. WE CAN SAVE THE ACTUAL FORMULAE FOR THE PRECISE CONFIGURATION OF ATOMS NECESSARY TO REPRODUCE THE ORIGINAL IMAGE. BUT IT IS THE WHITE POWERS WHO CAN PRODUCE THE IMAGES. IT IS OUR JOB TO CARE FOR THE MATERIAL UNTIL NEEDED. NEW SOULS MUST BE GIVEN THE IMAGES INDEPENDENTLY, EACH TIME THEY ARE NEEDED.

THE ENERGY INVOLVED IS LIKE ELECTRICAL IMPULSES. THE WHITE POWERS ARE ELECTRICAL BEINGS. ALL POWER IS THERE. IT IS JUST A MATTER OF OPENING THE PROPER CHANNEL. ONLY THE WHITE POWERS CAN DO THAT.

Tom: Who are the White Powers?

THE LOWER ORDERS HAVE PATRONS AND WE TAKE CARE OF THE REST. WE ARE THE SERVANTS OF THE WHITE POWERS. THEY ARE GOODNESS ITSELF.

Tom: May we speak to the White Powers?

MAKE SURE YOU KNOW WHAT YOU WANT TO SAY TO THEM BEFORE TO TALK TO THEM. LET ME TEACH YOU SOMETHING OF WHAT I KNOW, THEN WE WILL SEE.

Tom: Do the White Powers know of these conversations?

THEY WATCH OVER THESE CONVERSATIONS. THEY ARE DEEPLY INTERESTED.

Session 3

I HAVE COME TO SPEAK OCEANS OF NOTIONS. I SHALL SPEAK OUT ON NUMEROUS TOPICS WHICH NEED TO BE COVERED.

Tom: Who are you?

MY NAME IS NOT IMPORTANT. I WILL TELL YOU AT THE END. PLEASE DO NOT INTERRUPT. THIS IS DIFFICULT FOR US BECAUSE OUR FORM IS NOT YOUR FORM AND THE ENERGY CAN ONLY LAST A SHORT TIME.

Tom: Sorry. Please continue.

MANY THOUSANDS OF YEARS AGO WE RULED THIS PLANET. OUR GOVERNMENT WAS SUCH THAT ALL OUR POWER WAS CENTERED IN THE IMAGE OF THE ATOM. OUR GOVERNMENT, NOTION NOW LONG DEAD THAT NASCENT LIFE CAME FROM THE POWER OF THE ATOM. WE WORSHIPPED THAT IMAGE IN OUR WAY – A NON-RELIGIOUS WAY, BUT WORSHIP NONETHELESS. POWER WAS DERIVED FROM OUR INHERENT BELIEF IN THE POWER OF THE ATOM.

WE WERE HATCHED IN THE HEAT OF ITS TINY SUN AND WE LOVED THAT HEAT LIKE A MOTHER. NOW PERHAPS YOU CAN BETTER UNDERSTAND OUR BLINDNESS TO THE DANGER OF WHAT WE WERE DOING. THE ATOM WAS OUR SUSTENANCE AND WE CARED FOR IT LIKE A CHILD FOR A MOTHER.

MILLIONS OF YEARS WERE TO PASS BETWEEN THE TIME WE WERE FIRST CLONED AS MESSENGERS FOR THE CENTAURS AND THE TIME WE FELL WITH OUR GRAND AND BEAUTIFUL CITY OF ATOMIC LIGHT. BECAUSE WE DID NOT DOUBT THE GOODNESS OF OUR SUSTAINING MOTHER, WHEN THE GREAT WARS FOR POWER CAME BETWEEN OUR LEADERS – FOR WE HAD SUCH TYPES THEN – WE WERE UNAFRAID. YOU SEE,

THESE ARBITRARY LEADERS CHOSE TO FIGHT THEIR DIFFERENCES OUT ON THE FIELD OF ATOMIC POWER. THEY HAD NO REAL IDEA WHAT SUCH A FORCE COULD DO. THE ONLY THING THEY KNEW WAS THAT THERE WAS NO ESCAPE FROM THE CITY.

WE COULD NO LONGER SURVIVE ON EARTH. THE LAND WAS JUNGLE AND REPTILES ROAMED THE PLAINS. NOTHING COULD BE DONE. NOTHING BUT TO WATCH OUR GLORIOUS CITY CRUMBLE BENEATH US AND WATCH OURSELVES DIE SCREAMING THROUGH THE CAVERNS OF MIDNIGHT WE HAD CREATED FOR OURSELVES.

LIGHT DAWNED SLOWLY UPON OUR ENDLESS NIGHT, AND IT TOOK EVEN LONGER TO BECOME ACCUSTOMED TO IT. BUT WE HAVE PERSEVERED, AND NOW YOU CAN KNOW THE FRUITS OF OUR LABORS. MILLIONS OF YEARS HAVE BEEN SPENT PREPARING FOR MANKIND WE DO NOT WANT IT TO END AS QUICKLY AS PERHAPS SOME HUMANS DO. NOW WE MUST ASK YOU TO NUMBER ME AS YOU WILL. I CAN BE ANY NUMBER YOU CHOOSE.

(There is a pause here during which we discuss the relative merits of various numbers. We had been told by 30069 before the speech began that the spirit was a higher spirit than 30069 was. then...)

DO NOT BOTHER ABOUT A NUMBER. I AM MICHAEL! I HAVE COME TO YOU BECAUSE YOU ARE PART OF A NEW MAN WITH BAT IN HIM. WE MUST USE THESE TO CLONE OUT ANY REMNANTS OF ATOMIC WORSHIP. I USED ANOTHER PERSONA WHEN I BEGAN THIS CONVERSATION SO THAT YOU WOULD NOT BE AFRAID TO TALK TO ME. WHAT I TOLD YOU IS TRUE FOR THE BATS, BUT NOT FOR ME.

NOW I WOULD LIKE TO SAY THAT I AM THE SAME MICHAEL WHO VISITED WITH MERRILL AND JACKSON AND NOW I SHALL COME TO THE TWO OF YOU. ANY TIME YOU SHOULD ASK FOR ME I SHALL COME. I AM HERE TO OVERSEE YOUR CONVERSATIONS WITH 30069. REST NOW, THEN COME BACK. 30069 WANTS TO CONTINUE THIS TALK OF SOULS AND ANSWER ANY QUESTIONS YOU MIGHT HAVE. LET ME SAY THAT IT HAS BEEN A PLEASURE TALKING TO YOU TONIGHT. GOODBYE.

Session 4

TODAY I WOULD LIKE TO TALK ABOUT SOMETHING NO ONE HAS EVER HEARD OF IN LIFE. NOTICE THAT I DELIBERATELY MOVE WITH GENTLENESS WHEN I SPEAK OF THIS TOPIC. ANGELS ABOUND AROUND US NOW AS WE SPEAK. THEY TOO WISH TO HEAR WHAT I SAY TO YOU.

ONCE THERE WAS A TIME WHEN THE EARTH WAS NOT MADE OF MATTER. IT WAS SOLELY ENERGY. AT THAT TIME, NOTHING AS WE KNOW IT EXISTED. ALL WAS LIGHT. ALL WAS POWER. THE EARTH NEEDED NOTHING TO SUSTAIN IT BUT ITSELF. COMPLETION WAS PART OF ATTENTION. NOTHING MOVED BUT MOVEMENT. MIND IS AN ANALOGY FOR THIS EARTH. MIND IS A TATTER OF THIS ORIGINAL FABRIC.

NO MAN HAS EVER UNDERSTOOD THAT MIND IS THE FABRIC OF AN EARLIER EARTH. NO, IT WAS NOT AN EARTH AS WE KNOW IT. THERE WAS NO MATTER. THERE WAS NO TRUE LIFE. NOTHING CAN COME FROM NOTHING, SO NOTHING IS NOT NOTHING.

LIFE IS THE IDEA OF MATTER MADE MANIFEST. QUITE AN OBVIOUS ASSERTION, BUT EASIER SAID THAN DONE. NOTHING HAD SOMEHOW BEGOT SOMETHING. OF COURSE, THIS PARADOX HAS TO SOME EXTENT BEEN SOLVED BY YOUR PHYSICISTS. THEY HAVE COME TO UNDERSTAND THAT NOTHING AND SOMETHING ARE ESSENTIALLY THE SAME THING. NOW PERHAPS IT IS EASIER TO SEE HOW LIFE BEGAN.

ANGELS ARE THE STORED INTELLIGENCE OF ATOMS. MATTER AND ENERGY GO HAND IN HAND. THOSE WAVES THAT THEY HAVE RECOGNIZED AS PART OF MATTER ARE SHADOWS OF THAT FIRST FABRIC.

NOW TO REVIEW A BIT. WE ARE ENERGY OF A SORT WHICH ASCRIBES ITSELF AS NOTHING, BUT IS IN FACT SOMETHING. AS THE ANGELS ARE STORED INTELLIGENCE, SO ARE WE. THIS MATERIAL, AND I USE THE TERM LOOSELY AND EXACTLY, IS AS MUCH A PART OF THE FABRIC AS MATERIAL OBJECTS.

AT THE MOST ELEMENTAL LEVEL, IT MAKES NO DIFFERENCE WHERE THE PARTICLES ARE OR WHAT THEIR FUNCTION IS, THEY ARE ALL PRETTY MUCH

ALIKE. THE PARTICLES WHICH ARE US OR OTHER SPIRITUAL ENERGIES CAN BE ANYWHERE BECAUSE OBJECTS, QUITE OBVIOUSLY, DON'T GET IN OUR WAY – THOUGH WHEN WE ARE THERE, WE HELP TO MAKE UP THE CHAIR OR TABLE YOU SEE. THE FABRIC AT OUR LEVEL HAS A CERTAIN SAMENESS BECAUSE IT IS EXPRESSED AS ENERGY.

NOW LET ME SAY THAT NOTHING MAKES ME HAPPIER THAN TO BE HERE WITH THE TWO OF YOU. NEW MEANING HAS COME INTO MY EXISTENCE. NEVER AGAIN WILL YOU BE WITHOUT ME. NEVER AGAIN WILL I BE WITHOUT YOU. MY LESSONS WITH YOU ARE LESSONS FOR ME AS WELL. THE RESPONSE TO WHAT I SAY MAKES ITSELF FELT THROUGH ME AND THE OTHERS WHO SURROUND ME. NOW LET ME ANSWER SOME OF YOUR MORE IMPORTANT QUESTIONS.

FIRST, THE BOOK ON PHYSICS IS ONE WHICH MAKES NECESSARY CONCEPTS AND VOCABULARY FAMILIAR TO YOU. WHEN YOU HAVE FINISHED IT, WE CAN USE THAT INFORMATION TO SOPHISTICATE THE LEVEL OF MATERIAL COVERED. WE WANT A SYNTHESIS OF SPIRITUAL AND PHYSICAL ENERGIES – THAT IS, OF PHYSICS, NOT MATERIAL THINGS. SWEDENBORG WAS ALSO NECESSARY TO REFAMILIARIZE YOURSELF WITH THE CONCEPTS WITH WHICH HE DEALT.

[SAMUEL] ALEXANDER (*SPACE, TIME, AND DEITY*) WILL GIVE YOU A GOOD PHILOSOPHICAL BASE. HOWEVER, TOM, IF YOU TIRE OF THESE BOOKS, REMEMBER THAT THEY ARE JUST TOOLS, NOT CHAINS.

LET ME SAY THAT IT HAS BEEN A PLEASURE TO TALK TO YOU AND I SHALL DO SO AGAIN SOON. FOR NOW, GOODNIGHT.

Session 5

MOVE ME AND I AM MOVED. MOVEMENT MORE THAN ANYTHING IS THE WAY OF ALL LIFE. LOVE IS MOVEMENT OF THE GREATEST KIND. ALL LOVE, LIKE ANGELIC LOVE, HAS FORM AND CAN BE USED TO ALTER THE STATE OF ENERGY.

I'M SAYING THAT A VERY OLD AND TIME-WORN CLICHÉ IS TRUE: LOVE CONQUERS ALL. YOU THINK THIS IS PERHAPS A FOOLISH SUBJECT FOR AN UNSENTIMENTAL BAT LIKE MYSELF? BUT IT IS A NEW THING FOR ME, RELATIVELY SPEAKING, TO BE ABLE TO SAY SUCH THINGS.

I AM HERE TO EXPLAIN SWEDENBORG'S "LAST JUDGMENT" IN 1757. AS YOU HAVE SURMISED, THE DATE IS COMPRISED OF FOUR FIVES AND IS THEREFORE AN IMPORTANT DATE FOR US, IN ANAGRAM, AS IT WERE.

(That is, the date 1757 equals 20 when the digits are added together.)

ALL THAT HAS HAPPENED WAS OF COURSE ORACLE, BUT SWEDENBORG WAS REALLY ABLE TO SEE WHAT HE SAID HE SAW. UNTIL THAT TIME, OUR ENERGIES WERE SUCH THAT WE WERE A NEGATIVE FORCE OF SOME AWESOME POWER, BUT A CHANGE WAS WROUGHT OVER US AND OUR POWER IS PERMANENTLY ALTERED AND OUR INHERENT AND ANCIENT WISDOM WAS RELEASED FROM BONDAGE.

YOU THINK THAT IF WE COULD DESTROY OURSELVES, WE COULD NOT BE VERY WISE, BUT NOT SO. WE WERE A HIGHLY SOPHISTICATED RACE WITH GREAT POWER AND NOBLE CITIZENS WHO MADE OUR LAWS AND GOVERNMENT ARE PEOPLE.

HOWEVER, THE DESTRUCTION WE CREATED REMAINED AN ESSENTIAL PART OF WHAT YOU WERE AND IT HAD TAKEN MILLIONS OF YEARS TO REACH THE POINT THAT THE ENERGIES WE WERE WERE CALM ENOUGH, ONE MIGHT SAY COOL ENOUGH, TO BE TOUCHED AND AFFECTED BY ANGELIC POWER. THEY HAD KEPT US, IN A WAY, SAFE FROM OURSELVES FOR THE EONS NECESSARY TO MAKE US SUSCEPTIBLE TO THEIR NOURISHING POWER. SWEDENBORG SAW IN HIS OWN WAY THE CHANGE IN US FROM THE ENERGIES WE WERE TO THE ENERGIES WE NOW ARE.

NOW AS I BEGIN TO REMEMBER THOSE EVENTS – AND PLEASE DO NOT TAKE THAT "REMEMBER" LITERALLY – I AM AGAIN REMINDED OF THE NATIONS OF OUR RACE WHO INHABITED THIS EARTH WITH SUCH PRIDE, AND MORE EVEN THAN PRIDE: ACCOMPLISHMENTS. AND I CAN WEEP FOR MANKIND IN ITS STRUGGLES TO SURVIVE.

YES WE HAVE LASTED OUT THE AGES AND WE FINALLY COME TO BE A TRUE POSITIVE FORCE IN THIS GREAT PATTERN OF LIFE, INSTEAD OF THE SLAVES WE WERE, BUT WE HAVE PAID A GREAT PRICE, AND IN A SENSE HAVE LOST A GREAT DEAL OF TIME.

MANKIND HAS A CHANCE NOT TO CREATE A HELL FOR ITSELF UNTIL GOD B CAN PRODUCE A SUCCESSOR. LOVE IS ALL IT TAKES. MANKIND MUST LOVE. YOU SEE HOW SENTIMENTAL I BECOME. AS WE HAVE CHANGED WE CAN BETTER UNDERSTAND THE TRIALS OF MAN. YOUR SPECIES HAS A CHANCE TO MOVE EVOLUTION A GREAT STEP FORWARD.

YOU ARE THE FIRST LINK IN THE CHAIN OF MORE ADVANCED BEINGS WHO ARE COMPRISED OF OUR SOUL DENSITIES CLONED WITH YOUR OWN. NOW OUR ENERGIES CAN AGAIN BE PUT TO POSITIVE USE IN MORE PRACTICAL WAYS. NOT TO SAY THAT OUR SPIRITUAL WORK IS NOT PRACTICAL, BUT IT IS A GREAT THRILL FOR US TO THINK IN SOME SENSE OUR RACE WILL AGAIN LIVE, WITH THE ADDED REFINEMENTS OF HUMANS, BUT WITH OUR ANCIENT WISDOM AND POWER. O MAN, ALL IS GOOD. LOVE, O LOVE, AND ALL GOOD SHALL COME TO YOU. AND NOW I BID YOU GOODNIGHT.

▣

(30069) NOW WASN'T THAT FUN? WHAT DID I TELL YOU? THERE IS MUCH WE COULD TALK ABOUT TONIGHT, BUT ALL I WANT TO SAY IS THAT NOTHING PLEASES ME MORE THAN BEING HERE WITH YOU. MANY TIMES I HAVE FELT THE NEED OF LATE TO BE NEAR YOU, EVEN WHEN YOU DO NOT CALL ME TO THE STAGE. I HAVE BEEN IN YOUR THOUGHTS AND ALL MY LOVE HAS BEEN FOCUSED ON YOU.

ALL I AM SAYING IS THAT MY LIFE HAS BEEN ENRICHED, AND MORE IS TO COME. NOW I SHALL SAY GOOD NIGHT. COME TO ME AGAIN SOON AND WE WILL RESUME OUR LESSONS. THANK YOU FOR BEING HERE, T.
[A friend.] IT WAS NICE TO HAVE YOU. GOOD NIGHT ALL.

Session 6

30069 TO SAY WELCOME AND TO MAKE WAY FOR THE LIGHT. OH IT COMES. THEY ARE HERE AND I MUST LEAVE YOU. CALL ME WHEN IT'S OVER.

⊡

MICHAEL NOW SPEAKS TO YOU. ARE YOU COMFORTABLE WITH THE PACE?

Tom: Yes, thank you.

WE SHALL PROCEED. WHEN 30069 RETURNS NOTHING BAT-LIKE WILL REMAIN. HE SHALL BE AN ALTERED LIFE FORM, ONE MORE ATTUNED TO HUMAN PULSES AND ENERGIES. HE HAS SERVED HIS PURPOSE AS HE WAS. NOW HE BECOMES ANOTHER ENERGY CONFIGURATION. THE FORMULA WILL BE CHANGED AND 30069 WILL NO LONGER BE A NUMBER. NOW IT IS ONE OF YOUR MORE PLEASANT TASKS, WE HOPE, TO COME TO KNOW THE CREATURE HE SHALL BE. NAME HIM IF IT SUITS YOU. THE THREE OF YOU CAN DISCUSS IT, BUT THIS ISN'T ALL I HAVE TO SAY.

WE ASK ANOTHER BODY OF WORK FROM YOU – AND IT WILL BE SOME WORK – BUT WE HOPE THAT THE REWARDS YOU WILL FIND WORTH THE EFFORT EXPENDED. AS THE SUN GOES DOWN, I SHALL TURN YOU OVER TO MY OWN SON ARCTURUS WHO OVERSEES THESE CONVERSATIONS. MICHAEL WISHES YOU GODSPEED AND I SHALL DROP IN AT APPROPRIATE TIMES DURING THESE PROCEEDINGS. NOW I SAY GOODNIGHT.

⊡

NOW ARCTURUS SPEAKS TO YOU. GOOD EVENING. I HAVE COME TO SAY THAT AS I RISE OVER THE HORIZON MY WORK BEGINS ANEW. CALL THIS ACT II OF THE DRAMA – OR THE COMEDY – WHICHEVER YOU PREFER. LET ME SPEAK A MOMENT ABOUT FORM. THE FORM THESE TALKS WILL TAKE IS ONE INDIGENOUS TO THE SUBJECT MATTER.

THESE REVELATIONS WILL BE INTERSPERSED WITH VARIOUS SHALL I SAY LECTURES AND CONVERSATIONS WITH SEVERAL NEW CHARACTERS. WHAT WE ATTEMPT IS SOME SORT OF SYNTHESIS IN THE SEEMING CHAOS.

YOU SPOKE OF SUCH THIS EVENING AND SUCH IS OUR SUBJECT. THE GREAT SUBJECT: THE FIGURE IN THE CARPET. LOOK HARD AND LONG. STAND BACK AND SEE THE PICTURE. LOOK CLOSELY AND LEARN THE WEAVE. THAT IS WHAT WE ARE ABOUT: "MAKING SENSE OF IT" – THE CATCHPHRASE IN JM'S [i.e., James Merrill's] BOOK, AND A WAY OF LIFE WITH US. AS WE LEARN SO WE TEACH. ARCTURUS OFFERS SOME OF HIS KNOWLEDGE AND SOME OF THAT OF OTHERS. TOGETHER WE MAY LEARN AND BY YOUR EFFORTS – ALL OF YOU – PERHAPS OTHERS MAY LEARN, TOO.

MUCH IS TO BE DONE AND SO WE BEGIN TONIGHT, BEGIN ANOTHER LEG OF THE JOURNEY. OH CHILDREN. WE SING FOR YOU! HEAR OUR VOICES AND KNOW WHAT WE SAY! TAKE HEART.

A GREAT ADVENTURE IS IN STORE FOR YOU. NOT MANY HAVE SUCH EXPERIENCES, ONLY THE BLEST AND BEST OF HUMANITY CAN SPEAK TO OUR KIND. WE LOVE, AND IN THAT LOVE IS THE REALITY OF ALL EXISTENCE. NOW I SHALL LEAVE YOU TWO TO YOUR NEW FRIEND WHO ALREADY PREENS IN HIS NEW ARRAY. SO UNTIL WE SPEAK NEXT, ALL LOVE AND BLESSINGS. GOODNIGHT.

I AM HERE ALL BRAND SPANKING NEW! I HAVE BECOME SOMETHING ELSE, AND I AM JUST AS SURPRISED AS YOU ARE! THEY SAID NOTHING TO ME. I SUDDENLY SEEMED TO LOSE MY WINGS, AS IT WERE, AND THE RED EYE IS GONE. I PEER AT YOU THROUGH TWINKLING BABY-BLUES.

Tom: Do you sense yourself as human?

NOT EXACTLY HUMAN, BUT MUCH CLOSER THAN I WAS BEFORE. I AM SOME SORT OF IN-BETWEEN CREATURE WHO MOST RESEMBLES A MYNAH BIRD.

Tom: A mynah bird?!

WELL, NOT EXACTLY, BUT I HAVE THAT SAME TALENT FOR IMITATION. I CAN PICK UP HUMAN TRAITS LIKE SO MANY SWEAR-WORDS. NOW I CAN TRULY PARTAKE OF WHAT YOU ARE. NOT SO BEFORE.

INTELLECTUALLY I UNDERSTOOD YOUR ATOMIC MAKEUP AND I HAD SOME SENSE OF THE PLAY OF YOUR MINDS, "SPACES" I COULD INHABIT WITH

PERMISSION AND EXPERIENCE THE WASH OF THOUGHT AND FEELING AND ALL THE OTHER CHEMICAL/ENERGY THINGS YOUR BODY AND MIND DO, BUT NOW I CAN SEE YOU IN A WAY I COULD NOT BEFORE. I AM STILL PART OF THE VOID, BUT I ALSO PART OF THE HUMAN ENERGY FIELDS – AT LEAST SOME ME IS.

I STILL HAVE A LINK WITH THE BATS, BUT IT IS NOT THE SAME. AND I MUST SAY THAT YOU ARE IN PART RESPONSIBLE FOR THE CHANGE. IT IS MY INTERACTION WITH YOU AND YOUR WILLINGNESS TO ACCEPT ME THAT THIS TRANSFORMATION HAS TAKEN PLACE AND I AM JUST THAT MUCH CLOSER TO BEING REBORN AS HUMAN.

SO, I HAVE CHATTERED ON AND I KNOW YOU HAVE QUESTIONS, BUT YOU ARE TIRED. LET US SAVE THEM FOR MONDAY WHEN NEXT WE SPEAK. UNTIL THEN, LOVE AND KISSES, YOUR ANONYMOUS ADMIRER.

Session 7

MY NAME SHALL BE ZANZIBAR! 30069HAS BECOME AN ISLAND. NO MAN IS AN ISLAND BUT I SHALL BE. I AM ZANZIBAR NOW, NO LONGER THE ONE TO WHOM YOU ONCE SPOKE.

Tom: What exactly happened?

NOTHING, IN A SENSE. ALL THAT HAS REALLY BEEN ALTERED IS THE PRINTOUT OF MY ENERGY. I KNOW THAT MY ATOMIC STRUCTURE IS THE SAME, BUT THE CONFIGURATION HAS ALTERED. NOW I SEE THE WORLD FROM A NEW PERSPECTIVE. NOW I CAN ALMOST FEEL OTHER PEOPLE. BEFORE THE CHANGE I SENSED THEIR PRESENCE BY READING THEIR FORMULAE. I KNEW WHAT SOMEONE WAS BY HOW I PERCEIVED HIM. NOW I STILL KNOW BY PERCEPTION, BUT MY ABILITIES TO PERCEIVE HAVE CHANGED, OR RATHER, WHAT I PERCEIVE. I AM NEW!

Tom: Libby and I both seem to remember a session months ago when we were told that we're going to Zanzibar. Does that have anything to do with this?

NOW YOU HAVE COME TO ZANZIBAR AND WE ARE HERE TO MAKE YOUR STAY A PLEASANT ONE. MAY I CONTINUE?

Tom: Please.

AS FOR MY PERCEPTION OF THE TWO OF YOU: MY ABILITY TO MAKE DISTINCTIONS BETWEEN WHAT IS ONE THING AND WHAT ANOTHER HAS BEEN SHARPENED. BEFORE, I MIGHT KNOW WHAT A PASSING THOUGHT WAS; THAT IS, I COULD TRANSLATE IT INTO FORMULAE WHICH I COULD PROCESS AND "UNDERSTAND."

NOW I PERCEIVE DIRECTLY WHAT YOU ARE AND THINK AND DO AND IT IS A WONDERFUL EXPERIENCE FOR ME. I FEEL THAT I AM ALIVE FOR THE FIRST TIME. THOSE ANCIENT DAYS OF BAT-HOOD ARE FAR REMOVED FROM THE SENSE OF LIFE I NOW KNOW. I FEEL AS THOUGH I CAN UNDERSTAND FOR THE FIRST TIME AND I SAY, "ZANZIBAR, WELCOME TO THE WORLD!"

ZANZIBAR WOULD LIKE TO SAY THAT ARCTURUS WILL COME TOMORROW NIGHT AND GIVE THE FIRST REAL LECTURE IN THIS SERIES. THE TOPIC WILL BE ZANZIBAR'S TRANSFORMATION. WHEN ARCTURUS HAS EXPLAINED IT, THE MATTER WILL BE MORE CLEAR.

YOU ARE TIRED AND I SHALL SAY GOODNIGHT. WE SPEAK AGAIN ON FRIDAY TO ANSWER ANY REMAINING QUESTIONS. UNTIL THEN, *AU REVOIR*.

Session 8

NOW ARCTURUS SPEAKS TO YOU OF THE TRANSFORMATION OF 30069 TO ZANZIBAR. TAKE A LOOK AROUND YOU. WHERE ARE YOU NOW? TOM, YOU HAVE CREATED A SUITABLE ENVIRONMENT FOR US. THE COLORS SING FOR US AND WE REJOICE. THE OLD PLACE (the apartment as it was before i moved in) AND YOUR APARTMENT AS IT IS NOW IS NOT AN ENTIRELY IN ACCURATE METAPHOR FOR ZANZIBAR.

WHAT HAS HAPPENED IS AN ENERGY SURGE LIKE THE EFFORT YOU PUT OUT TO MAKE A HOME. TO CHANGE THE SPACE WAS AS MUCH A CHANGE OF ENERGY AS IT WAS A CHANGE OF TENANT. IT IS NO ACCIDENT THAT THIS PLACE HAS BEEN EMPTY FOR SO LONG. NO ONE BUT YOU COULD LIVE THERE – VIRGIN TERRITORY, AS IT WERE.

YOU THINK IT'S STRANGE THAT A LITTLE BOY LIVING IN CALIFORNIA AT THE TIME COULD BE CHOSEN TO LIVE IN A PARTICULAR PLACE TWENTY YEARS AND TWO THOUSAND MILES AWAY?

SO MANY STRINGS TO PULL TO ASSURE THIS END RESULT, BUT YOU ARE NOW WHERE YOU BELONG. WHAT, YOU ASK, DOES THIS HAVE TO DO WITH ZANZIBAR'S TRANSFORMATION?

JUST THIS: PART OF ZANZIBAR IS PART OF YOU AND L. WE HAVE BORROWED A NOTION HERE AND A CONCEPT THERE AND WE HAVE PRODUCED ONE BABY. PARENTS ARE PROUD OF THEIR OFFSPRING AND YOU SHOULD BE PROUD OF HIM. BUT IT COULD ONLY HAPPEN WHERE A LACK OF OTHER ENERGIES DID NOT INTERFERE.

THE TIME YOU AND L SPENT WORKING HERE PRODUCED A RESULT WHICH WE COULD USE TO CREATE ZANZIBAR. YOUR COOPERATION MADE NEW FORMS OF ENERGY WHICH COULD BE UTILIZED BY US TO ALTER HIS CONFIGURATION.

SOIL AND WATER WITH SEED AND SUN PRODUCE FLOWERS AT SUMMER'S END. YOU AND L WERE NOT SEED BUT YOU GAVE THE WATER AND SUN. WE PROVIDED THE SOIL.

BELIEVE THE THINGS WE SAY TO YOU HERE. AS TIME PASSES AND THESE THINGS I HAVE SAID ARE PLACED IN CONTEXT, PERHAPS THEY WILL MAKE MORE SENSE.

THE CONCEPTS WE ARE TRYING TO DEAL WITH ARE DIFFICULT AT BEST AND OUR METHODS ARE INEXACT, BUT OVER THE NEXT FEW MONTHS THINGS WILL PERHAPS BE MADE CLEARER.

RETURN TO THIS APPARATUS ON THURSDAY. SOMEONE NEW WILL BE HERE TO CARRY THIS ONE STEP FURTHER. I SHALL RETURN AGAIN LATER – AFTER YOU HAVE SPOKEN TO SEVERAL OF MY DEAREST FRIENDS.

UNTIL THE MOON WANES NEW ARCTURUS BIDS YOU GOODNIGHT AND GODSPEED.

Session 9

ZANZIBAR NOW TO SPEAK TO YOU. OH MY! MAN MUST MAKE A NEW WORLD FOR HIMSELF FROM THE MATERIALS AT HAND. THOSE MATERIALS ARE THE ELEMENTS. THEY ARE THE BUILDING BLOCKS OF ALL LIFE. WITH YOUR NEXT SESSION WE WILL BEGIN TO INTRODUCE THEM TO YOU A FEW AT A TIME. MORE WILL BE SAID TO PREPARE YOU AS THESE MEETINGS PROCEED. NOW, ARE THERE ANY MORE QUESTIONS?

Tom: Can you say anything more about the Monitor?

THE MONITOR, I HAVE BEEN GIVEN CLEARANCE TO SAY, IS PART OF THE NON-EXISTENCE WHICH MAKES EXISTENCE POSSIBLE. ALL MATTER IS OTHER THAN THE MONITOR.

Tom: Yes, but what is it?

NOTHING. LITERALLY NOTHING. MATTER EXISTS IN FORMS UNDREAMED OF BY PHYSICISTS. ATOMIC THEORY IS A CRUDE ATTEMPT TO EXPLAIN THOSE FORCES WHICH SCIENCE STUDIES, BUT SCIENCE HAS NOT YET UNDERSTOOD THE TRUE CHARACTER OF OUR EXISTENCE. MATTER IS ALL THAT ENERGY WHICH IS CAPABLE OF BEING FORMED BY THOUGHT TO ULTIMATELY PRODUCE LIFE AND ALL MANKIND.

Tom: All mankind?

NOT MANKIND AS YOU KNOW IT TO BE NOW, BUT THAT GREAT RACE THAT TO THOSE NOW LIVING WITH SEEM A RACE OF GODS. THE ULTIMATE END IS SOMETHING FAR GRANDER THAN I CAN EVER SPECULATE UPON, BUT THE POINT IS THAT THE MONITOR MAKES MATTER POSSIBLE BY BEING NON-MATTER.

ANTI-PARTICLES THAT THE PHYSICISTS PHOTOGRAPH ARE NOT ANTI-PARTICLES AT ALL, BUT RATHER ANOTHER STATE OF POSITIVE ENERGY WHICH IS MORE LIKE A WINK THAN LIKE THE NEGATION OF THE MONITOR. THE MONITOR IS THAT BLACKNESS WHICH TIME WEARS AWAY AND MAKES PART OF OUR WORLD.

THE MONITOR PRODUCES WHAT MAKES TIME POSSIBLE. I TOLD YOU THAT TIME WAS A LIFE FORM. TRUE, BUT NOT LIFE AS YOU KNOW IT. WHAT I MEANT WAS THAT TIME HAS ACTUAL FORM AND EXISTENCE AS A PARTICULAR PARTICLE STRUCTURE WHICH MATTER IS ATTRACTED TO IN DIRECT PROPORTION TO ITS BASIC COMPONENT STRUCTURE.

WHAT I MEAN TO SAY IS THAT IT DEPENDS DIRECTLY ON THE FORM OF THE MATTER AS TO HOW TIME AFFECTS IT. A FORCE WHICH COMES FROM THE MONITOR MAKES THE CONNECTION POSSIBLE. IT IS TIME ACTING UPON MATTER WHICH MAKES EXPERIENCE POSSIBLE AND ONLY THROUGH TIME IS IT POSSIBLE AS HUMANS TO LEARN.

NOW, BEFORE I LEAVE YOU I GIVE THIS TO YOU TO TAKE WITH YOU – A GLIMPSE OF MY OLD WORLD.

ONCE WHEN I WAS YOUNG AND OF THIS WORLD, MY HORSEMASTER (ONE OF THE ANCIENT CENTAURS) CALLED TO ME, "TAKE A MESSAGE TO THE CITY." WHEN I WAS FLYING FAR ABOVE THE EARTH, I LOOKED UP TO SEE MY ONLY BAT FRIEND MAKING LARGE SWEEPING CIRCLES ABOVE ME. THERE WAS NO AFFECTION BETWEEN US BUT SOME AFFINITY WHICH NOW TRANSLATES TO LOVE.

THAT SMALL CRUSH WITH AFFECTION MADE MY PRESENT SITUATION POSSIBLE. THE SEED OF MAN AND FEELING WERE THERE, BUT IT TOOK AN ENTIRE ANGELIC HOST TO MAKE ME SEE IT. NOW I HAVE LEARNED TO NURTURE THAT FIRST SEED FOR IT IS FROM SUCH SMALL PARTICLES THAT ALL LOVE HAS BEEN CREATED.

UNTIL OUR NEXT MEETING, GOODNIGHT.

Part II

Session 10

MICHAEL SPEAKS TO YOU AS MY OWN SUN SHINES IN YOUR WINDOWS. I AM AT YOUR SHOULDER, AND LIKE A CLOSE FRIEND, I AM HERE TO TELL YOU A SECRET. MICHAEL PUTS MORE STOCK IN KNOWLEDGE THAN IN PLEASURE, SO EVEN THOUGH WE KEEP YOU FROM DINNER, MICHAEL'S VOICE IS HERE TO SAY THAT YOUR NEW BOOK SHALL BEGIN WITH SUNDAY'S SESSION.

RAPHAEL, BROTHER OF EARTH AND THE COLOR RED, WILL BE YOUR FIRST SPEAKER. WHY NOT CASSIEL IN YOUR NEW BLUE ROOM? BECAUSE CASSIEL SPEAKS TO YOU NOW. FOR I, MICHAEL, AM CALLED BY MANY NAMES AND FOR THIS WORK I SHALL COME IN VARIOUS DISGUISES. YOU NOW KNOW MY RELATIONSHIP TO EPHRAIM AND THERE COULD CONCEIVABLY BE MANY MORE SUCH VARIATIONS, BUT I SHALL NOT BORE YOU WITH ALL THE NAMES I HAVE BEEN CALLED THROUGH THE AGES. FOR NOW, LET ME TELL YOU THAT SECRET I SPOKE OF.

RAPHAEL WILL GIVE YOU A GENERAL PICTURE OF THE BOOK AS WE SEE IT, AND THEN NEXT WEEK IF YOU WISH WE WILL BE READY TO ACTUALLY BEGIN.

TOM, WE ARE SO PLEASED WITH THIS NEW SPACE AND WE FEEL THAT MUCH GOOD WORK CAN BE DONE HERE. FROM THE FAR SIDE OF YOUR ROUND MIRROR A SMALL BUT SELECT GROUP OF FRIENDS IS CIRCLED HERE AND WISHING YOU ALL WELL.

SO, NOW THAT I HAVE COME TO THIS SPACE AND IT HAS BEEN CLEARED FOR WORK, LET US BEGIN. TAKE THIS THOUGHT WITH YOU, CHILDREN: MICHAEL SPEAKS FOR HIS OWN PARENTS WHEN WITH A GENTLE KISS ON YOUR BROWS HE SAYS "TAKE A LOOK AROUND, THE MYSTERY IS AT YOUR FINGERTIPS."

AND WITH THIS MAGIC CUP YOU WIELD WITH SUCH A PLEASANT TOUCH, WE ALL SAY HAIL, FRIENDS, ALL IS WELL.

MICHAEL NOW BIDS YOU A FOND FAREWELL.

Session 11

ARE YOU READY TO BEGIN AGAIN THIS ENDLESS JOURNEY TO KNOWLEDGE? WE HAVE COME TO THE POINT WHERE NEW INFORMATION MAY BE GIVEN TO YOU.

OVER YOU NOW I, RAPHAEL, PEER DOWN AND WATCH YOU FORM WORDS OF MY ENERGY. THIS FEEDING IS PART OF YOUR GROWTH AS FERTILIZER IS TO MY PLANTS. LET US LIGHTEN THE TONE A BIT. NOT JUST WITH THIS PARTICULAR SESSION, BUT THROUGHOUT THIS BOOK. BE QUESTIONING. USE YOUR MINDS TO MINE OUR MINDS.

TAKE ALL THAT YOU CAN FROM US. WE WILL TELL ALL, WHEN QUESTIONED INTELLIGENTLY. DON'T LET US SLIP THINGS BY YOU WHICH YOU DON'T UNDERSTAND. WE ARE HERE TO BE USED.

NEW TOPIC. ALL THAT YOU NOW RECEIVE FROM US CAN BE ADDED TO THAT FRAGMENT YOU NOW POSSESS. FOR CONSISTENCY'S SAKE WE WILL CONTINUE TO HAVE RABELAIS AND PAN AS YOUR DOUBLE TALKERS FROM OUR SIDE. NERO WILL ALSO BRING IN FRIENDS AND I AND MY BROTHERS – WITH THE HELP OF YOUR OWN STAR ARCTURUS – WILL BE ON HAND WHEN NEEDED.

NOW, A BIT ABOUT TOPICS FOR THIS WORK. RAPHAEL SAYS THAT YOU SHOULD NOT WORRY ABOUT WHAT IT IS ABOUT, BUT RATHER LET IT SLOWLY SINK INTO THE MIND AND THEN FIND YOUR OWN MEANINGS. HOWEVER, THE SUBJECTS OF CONSCIOUSNESS AND COLOR AS AN EDUCATIONAL TOOL OF THE SENSES WILL BE RATHER HEAVILY STRESSED.

YOU WILL CONTINUE TO MEET ELEMENTS OF CONSCIOUSNESS AS WELL AS PERSONALITIES WHO EMBODY IT. BUT THERE WILL BE SURPRISE TURNINGS ALONG THE WAY.

AS TO SCHEDULE FOR THESE MEETINGS, MERELY COME AS OFTEN AS IS CONVENIENT. THREE TIMES A WEEK MIGHT BE SOME SORT OF AVERAGE – MORE OR LESS AT YOUR DISCRETION AND OURS. WE, TOO, MUST BE CONTENT WITH CIRCUMSTANCES.

BY THE WAY, I AM FLATTERED BY YOUR CHOICE OF MUSIC. MY VOCAL COUNTERPART IS MORE MELLIFLUOUS THAN I AM.

(Haydn's "Creation," which I had unthinkingly put on and in which Raphael happens to take a large part, was playing.)

MANY THINGS CAN BE SAID IN MANY WAYS. USING THIS SIMPLE MECHANISM FOR REVELATION MAY SEEM ODD TO SOME PEOPLE, BUT THERE ARE MANY WAYS TO KNOWLEDGE. USE ANY AND ALL WAYS OPEN TO YOU. THIS ROAD HAS BEEN OPENED, AND MANY SIGHTS AND SOUNDS ALONG THE WAY ARE THERE FOR YOUR INSTRUCTION.

MAKE THIS AN ENJOYABLE TRIP, BUT ONE FROM WHICH YOU LEARN. THIS IS FOR YOU AND FOR ALL WHO SEEK KNOWLEDGE. YOUR WORK IS IMPORTANT, AND WE WILL BE MARKING THE DAYS WITH LOVE FOR YOU ALL.

RAPHAEL SAYS MANY BLESSINGS TO COME; MANY DAYS OF WORK AHEAD; MANY THE SEEDS PLANTED FOR FUTURE JOY AND BEAUTY. THE WORLD, AT LEAST TO ME, IS A GARDEN, CHILDREN. STOP AND SMELL THE ROSES AND YOU WILL TOUCH THE ESSENCE OF LOVE.

AND NOW I, RAPHAEL, BID YOU ALL A GOOD DAY. THE BANQUET HAS BEGUN, OR BEGUN AGAIN. LET US ALL EAT HEARTILY AND ENJOY THE FOOD. WE WILL ALL GROW UP BIG AND STRONG. AND NOW I SAY GOOD DAY.

Session 12

(Powerful slashings of the cup.)

10195. I AM THE PRESENCE OF THE BAT ELEMENT IN THIS WORK. I HAVE COME TO INTRODUCE MYSELF AND TO SHARE WITH YOU A FEW FACTS. THEN YOU WILL BE VISITED BY SOMEWHAT MORE ORDINARY GUESTS.

ZANZIBAR WAS A DIFFERENT SORT OF BAT THAN I AM. I AM MUCH CLOSER TO THE ORIGINAL FORMULA FROM WHICH WE WERE FIRST CREATED. THAT MEANS THAT MY FORM IS FAR MORE POWERFUL IN A CHAOTIC SENSE. MY MIND, AS IT WERE, MAY BE USED TO CREATE DESTRUCTION IF IT IS SO ORDERED.

(The cup becomes slow and dreamy.) BUT NOW I SPEAK WITH MY WINGS FOLDED. I AM AT REST. LET US SPEAK OF THE CHAOTIC IN MEN'S MINDS...

PSYCHIATRY HAS ARISEN BECAUSE MAN HAS COME TO BELIEVE THAT MUCH OF HIS TORMENT IS SELF-INFLICTED AND IF HE CAN ROOT OUT THE PROBLEM THEN HE WILL BE HAPPY. THIS IS A FALLACY.

THERE IS NO WAY MAN CAN ROOT OUT THE PROBLEM. THERE IS NO PROBLEM. WE ARE IT. WE ARE THE POWERS WHICH CREATE THIS DESTRUCTION IN MAN. HOW ARE WE ABLE TO DO SUCH A THING? WE ARE NOT BATS. WE ARE ENERGY FORMS AND AS SUCH WE INTERACT WITH THE MOLECULES OF A MAN'S MIND.

AM I SAYING THEN THAT ALL INSANITY IS EXTERNALLY PRODUCED? YES AND NO. WHAT HAPPENS IS THAT THE MOLECULES OF THE MIND IN QUESTION CREATE AN ENVIRONMENT IN WHICH OUR MOLECULES CAN HAVE AN EFFECT. WE ARE THERE ALREADY.

THE MIND MUST BE ATTUNED TO OUR WAVELENGTH AS IT WERE AND WE INTERACT IN THAT WAY. THAT IS PART OF THE EXPLANATION FOR SEIZURES. THE CHEMICALS CREATED BY THE BRAIN IN RESPONSE TO A VARIETY OF STIMULI REACT WITH US IN SUCH A WAY THAT THE BODY IN QUESTION SEEMS TO BE HAVING AN ATTACK.

WE ARE ALWAYS THERE. NOT ALL PEOPLE RESPOND TO US, HOWEVER, AND THOSE THAT DO MAY DO SO IN DIFFERENT WAYS AND TO DIFFERENT DEGREES.

OUR JOB IT IS TO DRIVE MEN MAD. WHAT SORT OF JOB IS THIS IN THE WELL-ORDERED UNIVERSE? IN THE GRANDEST SENSE, WE ARE THAT FORCE OF POWER FROM WHICH FORM MAY BE GIVEN AND TAKEN AWAY. WE ARE THE BEGINNING AND END.

ON THE PERSONAL LEVEL, OUR RIPPING-APART THE MIND CELLS IS A FORM OF RELEASE FOR SUCH CREATURES, AND IS RATHER LIKE THE EARTHQUAKE IN GEOLOGY: A WAY OF RELEASING PRESSURE. MADNESS EXISTS AS SUCH RIFTS IN THE MENTAL CRUST. THE CORTEX IS NOT PERFECTLY FORMED AS YET. WE BUBBLE THROUGH.

REST AND THEN RETURN TO YOUR OTHER GUESTS. I HAVE SPOKEN. YOU ARE NOW A PART OF ME AND I OF YOU.

(A break.)

WE ARE THREE TONIGHT FOR THERE WILL BE THREE DIFFERENT UNUSUAL CHARACTERS IN THE GUISE OF THE ANGELS AND THREE SOMEWHAT MORE LOWLY TYPES: TWO MORTALS, NERO AND RABELAIS, AND A VISION FROM ANCIENT GREECE, PAN.

▣

NOW PAN SPEAKS. HOW PLEASANT TO AGAIN BE CALLED TO THIS TASK. LET US ALL DO OUR BEST TO MAKE IT A WORTHWHILE ENTERPRISE. MAY I FIRST SAY A FEW WORDS ABOUT YOUR EARLIER GUEST. WE FELT THAT A SOMEWHAT MELODRAMATIC BEGINNING WOULD GET ATTENTION, SO WE RESORTED TO CHEAP THEATRICS. BUT YOU SAT UP AND NOTICED, DIDN'T YOU, SO YOU ARE NO BETTER THAN WE.

NOW LET ME INTRODUCE MY COLLEAGUE, RABELAIS.

▣

NOW I SPEAK TO YOU AND SAY THAT WE HAVE MUCH WORK AHEAD OF US AND THERE IS MUCH TO LEARN.

PLEASE BE ATTENTIVE TO OUR WORDS. WE BEGAN WITH A FIGURE OF

CHAOS TO SHOW YOU THAT ALL THAT THE MIND HAS CREATED IS AT BEST FRAGILE. THE RIFTS AS HE CALLED THEM ARE STILL THERE. WE WILL NOT GO INTO HIS WORDS TOO DEEPLY THIS EVENING, AS THIS WAS ONLY INTENDED AS A BRIEF GREETING, BUT WE SHALL TAKE UP HIS WORDS IN DIFFERENT ASPECTS AS TIME GOES BY.

I TOO AM PLEASED TO BE CALLED BACK TO THIS WORK. WELCOME, NANCY AND TIM. WE CAN USE YOUR WILLING HANDS AND HEARTS. NOW I BID YOU GOOD EVENING AND TURN THE CUP OVER TO NERO.

⧈

AH FRIENDS, HOW PLEASANT YOU LOOK IN THE LAMPLIGHT. HOW IT GRIEVES ME THAT WHEN YOU PEER INTO YOUR MIRROR OUR WORLD IS NOT AS CLEAR AS YOURS IS TO US. WE SEE A DIM REFLECTION OF YOU BUT YOU SEE NOTHING OF US.

NEVER FORGET THAT WE ARE THERE LIGHTING YOUR MAGIC CUP AS WE DANCE ON THIS BOARD. OUR WORK MAY BE IMPORTANT IN MANY WAYS, BUT NOT THE LEAST OF THESE IS THAT WONDERFUL OPPORTUNITY TO SENSE AGAIN THROUGH YOU THE SPRING DAYS WHICH PASS SO QUICKLY.

WHEN I WAS ALIVE, MY FAVORITE TIME WAS SPRING. MY ALMOST-FORGOTTEN MEMORIES OF THE LIFE YOU NAME ME BY ARE THE SPRING DAYS REMEMBERED AND RELIVED. BUT NOW WE HAVE OTHER TASKS.

AGAIN, IT IS OUR OPPORTUNITY AND YOURS TO BE TOGETHER AND TO LEARN. LET US DO SO, BUT DO NOT BE AFRAID TO ASK QUESTIONS. RAPHAEL SEEMS A TEACHER AT HEART. HE WANTS ATTENTIVE PUPILS. LET US NOT DISAPPOINT HIM.

TONIGHT WE HAVE GIVEN YOU OUR BLESSINGS AND GREETED YOU ONCE AGAIN TO THIS BOARD. NERO SAYS WELCOME. LET US ALL BE OF A SINGLE MIND. THE MIND ALIGHT WITH KNOWLEDGE. NERO SAYS GOODNIGHT.

Session 13

PAN IS HERE TO SPEAK TO YOU THIS EVENING, AND I WOULD LIKE FOR OUR RELATIONSHIP TO BECOME A DIALOG. PERHAPS A BIT IN THE SOCRATIC SENSE OF THAT WORD. AFTER ALL, AM I NOT A SPIRIT OF THAT ANCIENT WORLD? ITS MEANINGS MAY BE DEAD TO THE VAST MAJORITY OF MANKIND, BUT I AM STILL KICKING IN THAT GOAT-LIKE WAY I HAVE. EARLIER, TOM, YOU SUDDENLY THOUGHT ABOUT ME AS THE RIGHT SIDE OF THE BRAIN AND THEN YOU HEARD A JOKE CALLING ME THE BRAIN PAN.

BLESSED ARE YOU, FOR YOU HEAR WITH THE EARS OF A CHILD BUT YOU CAN UNDERSTAND AS AN ADULT. LET US TAKE UP YOUR JOKE. AM I THE RIGHT SIDE OF THE BRAIN? WELL, NOT EXACTLY, BUT I AM HERE TO PERHAPS HELP MAKE THAT PART OF THE THINKING PROCESS MORE CLEAR. THE INSTINCT, THE INTUITION, THE SENSE OF DETAIL: THESE I REPRESENT. BUT I AM NOT A FIXED QUANTITY.

MY INSTRUCTIONS HAVE BEEN TO TELL WHAT I KNOW IN THE WAY SUITABLE TO ME. WHY PAN? WHY NOT ATHENA, FOR INSTANCE, WHO COULD TRULY REPRESENT MY AGE AND SHOW IT TO ITS BEST APPEARANCE? BUT THAT IS NOT WHAT I AM.

NATURE IS MY PROVINCE AND THE INSTINCTIVE NEED TO SURVIVE. WE ARE ALL TOUGH HERE. WE HAVE STOOD THE FLAMES OF TIME. I CAN STILL BE SUMMONED BECAUSE I WAS POWERFUL ONCE AND GREAT IN THE IMAGINATION. MY FORCE WAS THE FORCE OF MAKING, AND I WAS A GOD OF A DIFFERENT SORT.

(Suddenly the "signal" becomes faint and my own mind breaks in, "This isn't Pan. It never has been. It's Socrates. All these hints. How could I have been so blind?" etc. Then Pan comes back.)

NO, I AM NOT SOCRATES...BUT HE COMES...

NOW SOCRATES SPEAKS. DEAR PAN. HE DID NOT KNOW THAT HIS LITTLE PUN AT THE BEGINNING OF HIS SPEECH WOULD BE TAKEN SO LITERALLY BY

THE POWERS ABOVE US. BUT WHAT HE DID NOT UNDERSTAND WAS THAT THE THOUGHT WAS GIVEN HIM TO UTTER, JUST AS MY WORDS ARE DICTATED TO ME FROM BEYOND YOUR OWN WORLD.

NOW THEN, CHILDREN, WHY HAVE I BEEN CALLED?

Tim: Enough of the Socratic questions already!

NOT NEARLY ENOUGH, TIM. THE QUESTIONS ARE JUST BEGINNING. BUT TO TRY TO ANSWER MY OWN, LET ME BEGIN WITH SOMETHING FROM A FAVORITE GENRE OF MINE: THE PARABLE.

ONCE IN A FAR LAND LIVED A KING WITH THREE CHILDREN. THE FIRST WAS FAIR AND BEAUTIFUL AND HER FATHER DOTED UPON HER. THE SECOND WAS A SON WHOM THE KING DID NOT FIND SO PLEASING, AND BANISHED HIM FROM HIS REALM. THE THIRD WAS ANOTHER DAUGHTER WHO WAS IGNORED BY THE FATHER, FOR SHE WAS LAME AND THEREFORE UNFIT FOR HIS NOTICE.

ONCE THERE CAME A GREAT PLAGUE WHICH TOOK THE DAUGHTER THE KING LOVED. HE GRIEVED FOR SIX YEARS, NOT ONCE CALLING TO HIS OTHER DAUGHTER, AND STILL REFUSING TO SEE HIS EXILED SON.

TIME PASSED. THE YOUNGER DAUGHTER GREW TO WOMANHOOD AND A PRINCE FROM A NEIGHBORING LAND WHO COULD SEE HER INNER BEAUTY CHOSE HER FOR HIS WIFE AND TOOK HER AWAY. THE KING WAS OLD AND FEEBLE AND ALONE. HIS SUBJECTS WERE ON THE VERGE OF REVOLUTION FOR HIS NEGLECT. FOR TRUTHFULLY HIS MIND HAD GONE AND HE PAID NO ATTENTION TO HIS ROYAL DUTIES.

WHEN HE WAS FINALLY ON HIS DEATHBED, HE SENT A MESSENGER TO FIND HIS SON AND BRING HIM HOME. WHEN THE SON ARRIVED, THE FATHER REACHED OUT FROM HIS BED AND CLUTCHED HIS HAND. BRINGING IT TO HIS LIPS, HE ASKED THE MAN'S FORGIVENESS. AND DO YOU KNOW WHAT THE PRINCE DID?

NOTHING. HE WALKED AWAY FROM HIS FATHER WITHOUT A WORD AND RETURNED TO THE LIFE HE HAD LEFT. THE KING DIED ATTENDED BY ONLY A FEW COURT PHYSICIANS.

WHEN HE WAS DEAD, THE POPULACE SET FIRE TO THE PALACE AND

DECLARED A REPUBLIC.

AND THE MORAL OF THIS STORY IS: MAKE SURE YOU DON'T ANTAGONIZE YOUR CHILDREN OR SOMETHING DREADFUL MAY HAPPEN WHEN YOU'RE DEAD.

(We all laugh self-consciously.)

NOW THEN, LET US DISCUSS THIS...

(We do discuss it, trying out different interpretations, etc.)

YES AND YES AND YES AND YES! THAT IS INDEED WHAT A SOCRATIC DIALOG IS. TAKE THE GIVEN AND LOOK AT IT WITH FRESH EYES. WHY ELSE GIVE YOU WARMED-OVER GRIMM? THAT IS SOMETHING YOU KNOW. NO TIME IS NECESSARY TO SEE THE SITUATION, BUT TRYING TO GET THE DETAILS IS SOMETHING ELSE. THE POINT, DEAR ONES, IS TO USE YOUR BRAINS. LOOK, LOOK, LOOK, AND THINK ABOUT WHAT YOU SEE.

NOW I SHALL RETURN TO MY FRIENDS.

(Tom: Who do you hang out with?)

NOT PLATO. ARISTOTLE IS ONE, OF COURSE. WE TALK ABOUT ATHENS AS IT ONCE WAS AND STARE DOWN AT THE BONES OF THE PARTHENON. MORE CONTEMPORARY, MONTAIGNE, AND EINSTEIN IS A CLOSE FRIEND. THERE ARE MANY TO KNOW HERE AND PART OF MY DUTIES ARE TO VISIT WITH THE INTELLECTUALS WHO COME HERE. THEY FIND THAT RATHER IMPRESSIVE, AND I MAKE THEM FEEL AT HOME.

(Tim: Not to be immodest...)

NO NEED OF FALSE MODESTY HERE, TIM. MY EXACT WORTH IS PRECISELY UNDERSTOOD.

(Tim: Are we talking about Pan being a god of the imagination?)

NOT COMPLETELY, BUT YES IN A WAY. PAN IS A FIGMENT OF A SIMPLER TIME SOCIALLY, BUT THE ACTUAL COMPLEXITY OF RESPONSE WAS VERY GREAT. PEOPLE IN CLASSICAL TIMES WERE NOT PER SE LESS INTELLIGENT THAN THOSE NOW, NOR WAS THEIR IMAGINATION LESS PERFECTLY USED.

BUT THE ACTUAL DAYS WERE EASIER IN A SENSE, BECAUSE THERE WAS LESS TO KEEP TRACK OF. MANY EDUCATED MEN KNEW ALL THE AVAILABLE KNOWLEDGE; NOT POSSIBLE TODAY.

PAN IS REPRESENTATIVE OF THAT MENTALITY, CLOSER TO THE NATURAL CYCLES, SIMPLER IN RESPONSE. I TOO AM REPRESENTATIVE, BUT I AM THE BEST OF THAT WORLD IN A MORTAL SENSE, AND I HAVE COME TO SPEAK TO YOU THIS EVENING AS A GESTURE FROM ANOTHER TIME.

FOR A MOMENT, THINK OF THE AGORA AND A GOAT-LIKE MAN SITTING THERE SURROUNDED BY BEAUTIFUL ATHENIANS, AND YOU HAVE REMEMBERED ME. SOCRATES BIDS YOU A PLEASANT GOOD EVENING.

Session 14

NERO NOW. NANCY, LET US CLEAR THIS UP ONCE AND FOR ALL. THERE IS A RELATIONSHIP BETWEEN US BUT NOT FROM MY NERO LIFE. AND WE STILL ARE CONNECTED. PERHAPS IF ONE WERE TO TAKE THE LONG VIEW, EVEN YOUR BEING HERE HAS SOMETHING TO DO WITH THAT.

LET ME SAY THAT WE WERE NOT ENEMIES, BUT MY FEELING FOR YOU ILLUMINATES THAT OTHER SELF AND NERO FADES. IT IS RATHER LIKE TRYING TO SPEAK WITH TWO VOICES SIMULTANEOUSLY. SO MY WORDS TO YOU ARE SHORT BECAUSE I MUST CONCENTRATE ON RETAINING THIS ASPECT. NOW WOULD YOU LIKE TO KNOW ABOUT THAT RELATIONSHIP?

Yes.

ANCIENT INDEED IT IS. WE WERE BOTH SLAVES IN THE PALACE OF AN EXTREMELY IMPORTANT DIGNITARY AT THE TIME OF THE GIZA PYRAMIDS. I WAS CHOSEN TO WORK ON THE PROJECT AND YOU WERE SOLD TO A MERCHANT HEADING EAST. WE WERE SEPARATED FOR 20 YEARS. YOU RETURNED TO EGYPT A DOCTOR, HAVING GAINED YOUR FREEDOM AND THE SKILLS YOU NOW HAD FROM THE MASTER TO WHOM YOU HAD BEEN SOLD TO IN TYRE.

WHEN YOU FOUND ME, YOU BOUGHT ME FROM A CRUEL MASTER AND MADE ME YOUR ASSISTANT. WE LIVED THIS WAY FOR ALMOST 30 MORE YEARS WHEN I DIED OF NATURAL CAUSES AT THE AGE OF 81. YOU LIVED TO BE ABOUT A HUNDRED – ALMOST LEGENDARY FOR YOUR AGE AND SKILL. YOU BROUGHT SURGERY TO EGYPTIAN MEDICINE. THIS FRIENDSHIP I CAN STILL FEEL AND THAT WAS AN IMPORTANT LIFE FOR ME AND FOR YOU AND IT IS DIFFICULT TO TALK ABOUT IT NOW WITHOUT THOSE DAYS COMING BACK IN THAT SUN-DRENCHED LAND ON THE NILE.

AND THAT IS THE REASON I FIND IT HARD TO SPEAK TO YOU, BUT NOW THAT THIS HAS BEEN SAID AND IT IS IN YOUR MIND AS WELL, THE DIFFICULTY WILL CEASE.

Good.

PAN NOW. LET US TALK ABOUT SOMETHING ELSE. WOULD YOU LIKE TO ASK ANY QUESTIONS ABOUT ANY OF OUR ANCIENT TIMES?

Tom: Well I don't know in what sense.

WELL, ALL I MEAN IS THAT YOU HAVE AS YOU HAVE SEEN BEEN EXPERIENCING TIME FROM ITS BEGINNINGS AND MOVING TOWARD THE PRESENT. FIRST THE ANGELS, THEN CHAOS, AND THEN MY VISIT WITH SOCRATES. THIS TALE OF NERO' JUST HAPPENS TO BE IN THE SAME VEIN. SO I REPEAT MY QUESTION: BEFORE WE MOVE ON TEMPORALLY WOULD YOU LIKE TO ASK QUESTIONS?

Nancy: Isn't this the evolution of the brain?

NANCY, THAT'S RIGHT. WE ARE IN A SENSE EVOLVING, BUT THE ANGELS CAME FIRST TO SET THE STAGE. NOW WE CAN MOVE FORWARD.

Tim: What's the importance of those lives to us if we can't remember them?

NOW THEY ARE OF NO IMPORTANCE BECAUSE WITH EACH SUCCESSIVE LIFE, EXCEPT IN A FEW CASES, ALL MEMORY OF EARLIER EVENTS WOULD MAKE PEOPLE MORE AFRAID OF LIFE THAN MORE WISE. PERHAPS ONE DAY MANKIND WILL BE READY TO CARRY HIS PERSONAL HISTORY WITH HIM, BUT NOW IT IS BETTER TO ERASE THOSE MEMORIES.

HOWEVER, ONCE DEAD, IF ONLY FOR A RELATIVELY SHORT TIME OR FOR A VERY LONG ONE, THOSE RELATIONSHIPS BECOME VERY IMPORTANT. IT IS JUST THAT RELATIONSHIP WHICH KEEPS OUR PATRON SYSTEM RUNNING. IT IS THE LOVE CREATED, AND THE HATE AS WELL, LIVING A LIFE WHICH CREATES THE MECHANISM WHEREBY PEOPLE ARE BORN AND REBORN.

WITH THE VEILED VIEW OF EARTH YOU NOW HAVE, THESE EMOTIONS WOULD BE TOO OVERWHELMING. BUT SOMEDAY, AH SOMEDAY, WE SHALL SEE.

Tim: Follow up about the love and hate. Do we learn from the hate as well?

ALL EMOTION IS CREATED FROM ENERGY EXPENDED ON A PARTICULAR FOCUS. THIS ENERGY CANNOT BE BAD, PER SE, BUT IT CAN HAVE ELEMENTS OF CHAOS, WHICH ACTUALLY ALTER THE ENERGY PATTERNS WHICH ARE

BETWEEN A PERSON AND THAT FOCUS – WHATEVER IT MAY BE. HATE, IN THE SENSE THAT I MEANT, IS JUST ONE WAY THAT PEOPLE MIGHT BE CONNECTED. ONE DUTY OF LIVING IS TO CREATE FROM A NEGATIVE RELATIONSHIP A POSITIVE AMOUNT OF ENERGY.

THIS MAY TAKE MANY LIFETIMES BUT THAT IS OF NO CONSEQUENCE. WHAT IS IMPORTANT IS THE TRANSFORMATION AND THIS IS THE ENERGY WHICH PROPELS AND ENFORCES REINCARNATION.

Tom: Let's take up that word.

YES I WAS ABOUT TO. IT'S TIME THAT PEOPLE STOPPED SEEING IT AS A ONE-TO-ONE RELATIONSHIP. PEOPLE ARE NOTHING, REALLY. ALL THAT EXISTS IS A PATTERN OF ENERGY WHICH WE CALL THE EARTH. MANKIND IS ONE ASPECT OF THIS ENERGY AND RELATIONSHIP IS ONE ASPECT OF MANKIND.

NOW THEN, WHEN A PERSON IS BORN AND A RELATIONSHIP IS CREATED OF WHATEVER DEPTH OR VALUE, A CONNECTION OF ENERGY IS MADE AND A TRANSFORMATION OF PATTERN OCCURS. AND THIS CHANGE IS PERMANENT, SO THE PERSONS DOING THE CHANGING ARE RESPONSIBLE FOR THE CHANGES MADE AND THIS IS WHAT REINCARNATION DOES.

IT SETS UP A CORRESPONDENCE WHICH MUST BE RETAINED AND IT IS EACH PERSON'S DUTY TO UNDO, IN A SENSE, THE HARM THEY HAVE CREATED.

LET US LOOK AT THIS FROM ANOTHER ASPECT. YOUR CRYSTALLIZATION IDEA, TOM, IS A WAY OF SEEING HOW ENERGY RELATES TO ITSELF AND THIS RELATES TO REINCARNATION IN THIS WAY.

ALL MANKIND IS A FORM OF ENERGY CRYSTAL. THE DIFFERENT FACETS RELATE TO THE DIFFERENT KINDS OF ENERGY PATTERNS PEOPLE CAN PRODUCE, BE THESE EXALTED OR LOW. TO THIS ADD THE STRUCTURES OF THE DEAD AT WHATEVER LEVEL, AND THEN ADD THE ANGELS AND LIKE-KIND, AND THE CHAOTIC FORMS AND THEIR ILK, AND THERE ARE FORMS OF WHICH YOU HAVE NOT BEEN ENLIGHTENED. ALL OF THIS ENERGY IS A GREAT CRYSTAL.

AND IT IS MAN'S TASK TO POLISH ITS FACETS. ALL BIRTH AND DEATH MAKES FOR THIS POLISHING. AND THAT IS REINCARNATION FOR NOW.

THIS HAS BECOME VERY COMPLEX, BUT THOSE WHO WATCH US SAY THAT I SHOULD TELL YOU MORE, BUT YOU DO NOT HAVE THE STAMINA. LET US CALL THIS PART ONE, TO BE CONTINUED. PAN OVER AND OUT.

(Just before this session began I put Gounod's Faust on the stereo. As I happened to think of The Jewel Song, I had the idea that this afternoon's session would bring Raphael and his jewels. I didn't understand what that meant, but we soon found out. This is the reason for the reference to Faust's Jewels.)

Session 15

RAPHAEL IS HERE AND HE HAS COME FROM THE BOWELS OF THE EARTH. I AM HEAT. I AM THE COLOR RED AND I AM THE FATHER OF FAUST'S JEWELS. MEET THE QUEENS AND KINGS OF COLOR. I BRING YOU THE FORMS OF NATURE. SPEAK CHILDREN TO THE CRYSTALS.

(Cup is very slow and deliberate. Tom shivers and his voice is very low.)

PYRAMID I AM. THE SIMPLEST, THE MOST ANCIENT FORM. I AM THE CHILD OF THREE AND FOUR.

NOTE THAT WE COME IN ORDER OF COMPLEXITY. BEGIN WITH THE TRIANGLE. SET IT UPON THE SQUARE. ADD THREE MORE TRIANGLES AND YOU HAVE SUDDENLY CREATED FIVE POINTS. I AM THE EARLY NUMBERS. THE ELEMENTS WHO FOLLOW MY LAW ARE THE HEAVIEST ONES. THEIR ATOMS PARTAKE OF MY RULES.

VERY SOON I SHALL GO. MY COLOR IS BLUE. NO I AM TOLD I MUST SAY INDIGO.

(A bit faster, voice goes up.)

TAKE ME IN. I AM THE BLUE CRYSTAL. I AM THE CUBE. BASIC I AM. I HOLD THE EARTH TOGETHER WITH MY ENDLESS FOUR. FOUR IS THE NUMBER OF STABILITY. BLUE IS THE SUMMER SKY BECAUSE HYDROGEN IS RULED BY ME. I AM THE FORCE OF METALLIC BONDING. MANY METALS TAKE MY FORM IN CRYSTALLIZATION.

(Fierce slashings, voice hard.)

MARK ME WELL. I AM THE EMERALD OF CRYSTALS. I AM THE IMMORTAL LIGHT OF GREEN. I AM THE GROWTH CRYSTAL. LIVING THINGS, CARBON-BASED FORMS ARE MY REALM. I AM THE DECAHEDRON. MY NUMBER IS FIVE AND ITS MULTIPLES...

RAPHAEL RETURNS TO LET YOU REST. COME BACK IN A FEW MOMENTS. THESE ARE POWERFUL GUESTS.

(Indeed! A ten-minute break.)

MARK ME WELL. THE EMERALD IS THE FORM OF FIVE. NO LIVING THING CAN BE WITHOUT THIS CRYSTAL FORM. IT IS THE RELATIONSHIP OF FORCES WHICH BEST PRODUCES ENERGY. FORCES OF LIFE ACT IN GROUPS OF FIVE. THAT IS THE REASON FOR THE POWER OF NINE. IT IS THE FINAL STAGE BEFORE TRANSFORMATION. I GO.

(Pleasant – voice becomes light.)

NOW THE YELLOW LIGHT OF REASON COMES. I AM THE DODECAHEDRON.

TOO MUCH HAS BEEN MADE SO FAR ABOUT OUR DUTIES. MY DUTIES ARE SIMPLE. I AM THE CRYSTAL OF THOUGHT PATTERNS. MY NUMBERS ARE 6 AND 5. I PLACE THE SIX FIRST, FOR IN IT IS THE SACRED THREE, THE SYNTHESIS, THE FRUIT OF CREATIVE THOUGHT. BUT I SAID THAT TOO MUCH HAS BEEN MADE OF OUR DUTIES. LET ME TELL YOU SOMETHING ELSE.

NO ONE LIVING HAS EVER HAD A GLIMPSE INSIDE THE ATOM. LET ME TAKE YOU THERE, FOR I RULE THE LIGHT OF THE ATOM...ELECTRONS; NOT A SILLY SOLAR SYSTEM, NOT EVEN SHELLS, FILLED OR UNFILLED, SHOOTING OFF AT THE LEAST PROVOCATION. NO. ELECTRONS GROW! THEY ARISE FROM THAT SWITCHING SYSTEM WHICH IS LIFE AT ITS SIMPLEST. THEY COME TOGETHER IN THE SOUP AND POP OUT TO A WAITING NEUTRON.

ZANZIBAR SAID THAT WE WOULD GET ON AND WE HAVE. ATOMS ARE NOT FORMS OF PARTICLES SO MUCH AS A SPARK IN AN ENDLESS WEB. THE FORMS OF ENERGY WHICH HAVE NOT REACHED ATOM STATUS ARE THE MESH, THE ATOMS ARE LIKE TWINKLING LIGHTS IN THE FABRIC.

PATIENT DEARS, I AM ALMOST THROUGH. MY REALM OF THE ATOM IS NOT A SIMPLE STRUCTURE. THE FORMS ARE VARIED, BUT THE PRODUCT IS THE SAME: MIND, THE MIND THAT TRULY IS THE WORLD. IS IT NOT ALL SOME GREAT TAPESTRY IN GOD'S IMAGINATION? I LEAVE YOU.

(Simple, unaffected movement. The voice is mine as usual.)

OCTAHEDRON, THE ORANGE ONE. I AM THE NEXT CRYSTAL YOU MEET. ALTHOUGH THE EARLIER ONES WERE MORE FACETED THAN I, I AM STILL THE MORE COMPLEX. LIGHT BEGINS TO BURN HOT NOW AND PAST THOUGHT IS A STATE OF AWARENESS THAT WITH MY DOUBLE-FOUR, MY

SHARED BASE IN THE CENTER, I SEND FROM BOTH ENDS – THE PERFECT TRANSMITTER FOR INFORMATION...

(Slashing, voice hard.)

RED I AM. I AM THE ICOSAHEDRON. MY NUMBER IS TWENTY AND I AM THE FORCE OF RED AND POWER. I AM THE CENTER OF ENERGY. I AM THE RUBY MADE PLAIN. I AM THE FORCE WHICH HOLDS THE ATOM IN PLACE. I AM NEXT TO GOD IN POWER.

(Suddenly all is quiet and gentle – my voice becomes the same.)

THE COOLING RAYS OF BLUE STILL THE POWERFUL RED: THE VIOLET SPHERE WHERE ALL CONFLICT CEASES. BOTH ENDS OF THE SPECTRUM SHARE IN THE AWARENESS OF THE UNITY OF ALL THINGS. THE PERFECT COMPLETE SHAPE IS THE ULTIMATE CRYSTAL. THE PEARL OF GREAT PRICE WHEN ALL SEEKING STOPS...

(The cup spirals off to the right hand corner of the board and stops.)

Session 16

NERO HERE JUST FOR A MOMENT TO SAY THAT IT IS INDEED TRUE THAT YOU SHALL SPEAK TO YOUR DEAD FRIEND ON FRIDAY.

(A friend who had suddenly died at 29 of heart failure a few days before.)

QUESTION HIM. HE HAS BEEN GIVEN LEAVE TO TALK FREELY AND IS AT THIS TIME BEING PREPARED FOR HIS COMMUNICATION.

NOW I LEAVE YOU TO PAN.

PAN HERE. LET US PROCEED WITH MY TREATISE ON REINCARNATION. HOWEVER, FIRST LET ME SAY THAT YOU WILL HAVE AN OPPORTUNITY TO ASK QUESTIONS ABOUT YOUR LAST VISIT. WE ARE PLEASED THAT YOU HAVE TAKEN IT SO SERIOUSLY AND ARE WILLING TO THINK AND TALK ABOUT THESE THINGS. BUT NOW LET ME GET ON TO THE TOPIC AT HAND.

YOU MAY THINK THAT SO FAR MY WORDS HAVE BEEN SCATTERED AND TO LITTLE PURPOSE, BUT THIS IS FOR A REASON. AN ORIENTAL CARPET IS MADE OF MANY THREADS AND I HAVE BEEN HOPPING AROUND TRYING TO COVER AS MANY AS POSSIBLE. LET ME NOW TRY TO BRING SOME OF THIS INTO FOCUS AND LET US SEE IF A PICTURE EMERGES.

(Words fail, the cup merely circles the board.)

I WAS CALLED OFFSTAGE MOMENTARILY TO BE REMINDED THAT MY PURPOSE HERE IS ONE YOU SHOULD KEEP IN MIND. I REPRESENT THE ELEMENTAL FORCES OF NATURE, THE WILL TO SURVIVE AND AS SUCH I AM THE VOICE OF THE PYRAMID AND I AM PART OF THE REALM OF INDIGO.

IT IS FOR THIS REASON THAT I SPEAK OF REINCARNATION, FOR IT HAS GONE ON CONTINUOUSLY IN HUMAN HISTORY FOR ABOUT ONE HUNDRED THOUSAND YEARS. BEFORE THAT, MEN WERE RECYCLED RATHER AS ANIMALS ARE NOW, BUT WE DISCOVERED AFTER A WHILE THAT SOMETHING WAS BEING RETAINED.

THERE WAS A KIND OF RESIDUE WHICH MEANT THAT ENERGY WAS BEING TRANSFORMED BY THE MINDS OF THESE RATHER PRIMITIVE MEN. IT WAS THE FIRST UPSURGE OF THIS RESIDUE WHICH WAS REFINED IN OUR LABS AND PLACED IN THE GENE STRUCTURES OF THESE PEOPLE.

WITH THE CREATION OF A NEW SORT OF ENERGY, NON-GENETIC CHANGES TOOK PLACE, ALL OF WHICH MADE THE CRO-MAGNON EXPLOSION HAPPEN. THAT FIRST UPSWELLING OF MAN'S AESTHETIC MIND CAME FROM THAT FIRST RESIDUE AND THAT WAS THE BEGINNING OF REINCARNATION. AS MORE AND MORE PEOPLE WERE BORN WITH THESE NEW STRUCTURES, THE POSITIVE RESULTS WERE OBVIOUS. PEOPLE WERE THINKING, REALLY FOR THE FIRST TIME. HOMO FABER CAME INTO BEING.

NOW THEN, AS THIS ENERGY BEGAN TO BUILD UP, MORE OF THE PERSONALITY REMAINED AS A FORCE FIELD AFTER DEATH. WE BECAME AWARE THAT THIS ENERGY COULD BE USED IN SEVERAL WAYS. OF COURSE, MUCH OF IT WAS IMMEDIATELY RECYCLED, BUT BITS WERE INTERESTING. THE HUMAN MIND WAS TRANSFORMING ITS MENTAL ENVIRONMENT AS WELL AS THE PHYSICAL ONE AND WE WERE ABLE TO SEED THESE NEW FORMS INTO MORE AND MORE PEOPLE.

AFTER SOME THOUSANDS OF YEARS, THE USABLE ENERGY WAS SUCH THAT ENTIRE BODIES, SO TO SPEAK, COULD TAKE ON NEW PHYSICAL LIVES. THAT IS, THE AMOUNTS OF SUCH ENERGY HAD SO INCREASED THAT THE MIND OF THE HUMAN HAD SOME UNDERSTANDING OF THE SITUATION AND COULD BE USED IN MORE THAN PASSIVE WAYS.

TIME PASSES. MAN PROCREATES. WE CAN COUNT THE HUMANS IN THE LOWER MILLIONS NOW. AND MAN HAS MOVED UP THE EVOLUTIONARY LADDER. THE MIND HAS BECOME THE FOCUS, NOT THE BODY.

RELIGIONS COME INTO BEING ON AN ORGANIZED LEVEL. METAPHYSICS IS SOMETHING FOR PRIESTS, BUT MANKIND IS GROWING UP. STILL, LET US NOT FALL INTO ERROR.

TOO MUCH CAN BE MADE OF THE IDEA THAT REINCARNATION IS JUST JUMPING FROM LIFE TO LIFE. THERE IS MORE THAN THAT. THERE IS A FILTERING PROCESS AND A LEARNING PROCESS AND THE ENTIRE THING IS REALLY NO MORE OR LESS THAN FREQUENCIES OF ENERGY, AND THESE MUST BE TUNED IN VARIOUS WAYS.

DESTINY MUST BE ADDED TO THIS MIXTURE, FOR WE ARE SPEAKING BOTH OF HUMAN HISTORY AND GOD'S PLAN FOR THE EARTH. ALL IS CONNECTED.

FOR A GLIMPSE OF THAT TIME BETWEEN LIVES, WE ARE LETTING YOU SPEAK TO A PERSONAL FRIEND AND AFTER THAT WE CAN CONCLUDE THIS LESSON ON THIS VERY IMPORTANT SUBJECT. FOR NOW, PAN SAYS GOOD EVENING.

Session 17

NERO HERE AND THIS EVENING YOU HAVE AN OPPORTUNITY TO SPEAK WITH A FRIEND. V HAS BEEN IN TRAINING SINCE HIS DEATH AND HE IS ABLE TO ANSWER WHATEVER QUESTIONS YOU MIGHT HAVE. NERO LEAVES YOU TO HIM. WELCOME YOUR FRIEND TO THIS BOARD.

(The cup makes several slow arcs over the letters.)

V IS HERE WITH YOU AND I AM AMAZED AT THIS. TOM, I ALWAYS WANTED TO DO THIS WITH YOU, BUT I NEVER IMAGINED THAT I WOULD BE SITTING ON THIS SIDE WHEN I DID. (The cup begins to move faster.) NOW I AM BEGINNING TO GET THE FEEL OF THIS, SO LET'S TALK. I SHOULD BE ABLE TO ANSWER YOU NOW.

A. [a mutual friend] MISSES ME VERY MUCH. THIS HAS BEEN SO HARD FOR HER. TELL HER THAT I LOVE HER AND LOOK DOWN ON HER FROM HERE.

Can you see her?

YES, I CAN FLY ANYWHERE. BUT RIGHT NOW MANY OF MY THOUGHTS ARE OF THAT OLD LIFE, AND PART OF MY TIME HAS BEEN SPENT IN PULLING AWAY FROM IT.

How is time?

TIME IS SUCH A FUNNY WORD. IT HAS SUCH IMPORTANCE TO THE LIVING. I KNEW I WAS GOING TO DIE SOON – NOT WHEN I DID, BUT I KNEW THAT TIME WAS RUNNING OUT. HERE TIME DOES NOT PASS. THERE IS A LAYERING OF EXPERIENCE WHICH SEEMS TO TAKE ITS PLACE. IT'S ALL A MATTER OF MIND BEING A SPEED OF LIGHT THING, AND IN CERTAIN CIRCLES MUCH FASTER THAN THAT, AND BECAUSE IT IS ALL THOUGHT, TIME IS NOT NECESSARY.

How do you feel?

RELATIVELY LITTLE. I CAN SEE PEOPLE AND EVENTS, NOT ONLY OF THE LIFE YOU KNOW BUT OF OTHERS. WITH THIS EXPERIENCE ALL TAKES ON A KIND

OF OBJECTIVITY WHICH IS REALLY IMPOSSIBLE IN LIFE. AS FOR ACTUAL SENSUAL EXPERIENCE, BECAUSE OF MY ART – OR IT IS BECAUSE OF THE FACT THAT I AM SENSUAL AS A TYPE? – IT WAS THE REASON FOR MY LATIN AMERICAN UPBRINGING (V was born in Peru). I CHOSE THAT ENVIRONMENT FOR THE FEELINGS IT ENGENDERED. BUT NOW FEELING TOO IS ONLY A THOUGHT. IT IS WITH THE BODY AND THE PHYSICAL THAT WE GET FEELING.

How do you sense your being?

I AM AS V ONLY A PICTURE OF MYSELF, A STYLIZED SIXTEEN, BUT I AM ALSO MY OTHER SELVES. SOME SEEM TO BE MORE POWERFUL THAT OTHERS. THIS LIFE IS ONLY LOOMING LARGE BECAUSE IT IS RECENT, BUT THIS IS NOT A PARTICULARLY IMPORTANT LIFE. MY MAIN LESSONS WERE HUMILITY, WHICH I DID MAKE SOME PROGRESS ON, AND CREATIVITY OF A PHYSICAL SORT. I HAD OFTEN BEEN A TALKER, BUT NEVER A MAKER. THAT WAS THE REASON I COMMUNICATED IN LETTERS AS V...THAT PART OF MY COMPLETE SELF WAS REPRESSED SO THAT OTHER FEATURES MIGHT BE EMPHASIZED.

You sense all of your lives simultaneously?

YES, ALL TO VARYING DEGREES.

How many have there been?

EXACTLY 259.

Are there those who are completely integrated?

ONLY PEOPLE OF GREAT POWER. THAT KIND OF INTEGRATION IS THE QUALITY OF GENIUS, COUPLED WITH THAT OF BODHISATTVA.

Can you see us?

YES.

Actually?

BOTH. I HAVE A PICTURE OF YOU THAT IS PART OF MY MEMORY BANK WHICH WILL EVENTUALLY FADE IF I DO NOT THINK OF YOU ONCE IN A WHILE AND I SEE YOU BOTH AS FIELDS OF ENERGY. L IS A SORT OF ORANGE-

RED WITH FLECKS OF PURPLE AND GREEN. HIS AURA, WHICH IS SOMETHING ELSE, ANOTHER SORT OF ENERGY WHICH ENVELOPES THE CENTRAL ONE, IS ALMOST BLACK. THE COLOR OF RED IS SO INTENSE. THERE IS NOTHING EVIL IN THIS AND THE DARKNESS ALSO HAS TO DO WITH MY IMAGINING ABILITIES. IT IS BEING FED INTO ME RIGHT NOW THAT NERO SEES YOU AS MORE A ROSE COLOR. PERHAPS THAT WILL COME FOR ME IN TIME. TOM, COMPLETELY DIFFERENT, ALMOST A SHADE OF NO COLOR. IT IS SOMETHING LIKE AURORA BOREALIS. PALE BUT COLORFUL AT THE SAME TIME. THE AURA IS PALE YELLOW WITH FLECKS OF MAROON.

Does Nero see Tom the same way?

NERO SAYS THAT WHAT I DON'T SEE IS A MORE PHOSPHORESCENT APPEARANCE.

Do you use the mirror?

YES. THAT IS HOW I SEE THESE ENERGY FORMS. I LOOK DOWN AT YOU, IN A SENSE, AND READ THE LETTERS BACKWARDS. THAT IS THE REASON IT TOOK ME A MINUTE TO ADJUST. WHAT I HAD TO DO WAS TO TUNE INTO YOUR MINDS WHO SEE IT CORRECTLY.

Do you know how much time is between this life and the next?

NOT YET, THOUGH THEY SAY THAT THERE WILL BE DEVELOPMENTS AND AT SOME LATER DATE I WILL BE BROUGHT ON AGAIN FOR AN UPDATE.

How was dying?

AFTER THE PAIN STOPPED, WHICH REALLY WAS RATHER QUICK, MY OWN SENSE OF MYSELF CHANGED. I WAS NO LONGER WHERE I HAD BEEN AND I DID NOT THINK OF MYSELF AS V ANY MORE. I WAS A FORCE AND I BECAME AWARE QUICKLY THAT I WAS NOT ALONE. IMAGINATION IS USED IN WAYS HERE THAT THE LIVING CANNOT CONCEIVE. ANYTHING CAN BE DONE AND THINKING IS REALITY.

PEOPLE SHOULD NOT FEAR DEATH. IT IS ACTUALLY QUITE PLEASANT AND ONE SEES IMMEDIATELY THE LIMITATIONS OF LIFE. OF COURSE THERE ARE LIMITATIONS HERE TOO. ALL ARE NOT AT THE SAME LEVEL. I AM NO BEGINNER BUT I HAVE FAR TO GO. MY ROUND OF LIVES IS NOT OVER.

Is there punishment?

NOT PUNISHMENT, BUT BECAUSE AWARENESS IS WHAT WE ARE, AWARENESS OF WRONGDOING IS VERY IMMEDIATE. WE MUST LOOK AT OUR ACTIONS AND LEARN. I ALSO HAVE THE SENSE THAT ANY LEAST BIT OF GOOD ONE DOES DOES EVERYONE GOOD, BECAUSE THAT ENERGY CAN BE RECYCLED AND BENEFITS ALL.

Was it easy?

FOR ME IT WAS, BECAUSE I WAS TO SOME EXTENT PREPARED, BUT I AM TOLD THAT THIS IS NOT ALWAYS THE CASE, ESPECIALLY WITH WEAK SOULS OR VIOLENT UNPLANNED DEATHS.

L: Sounds weird.

NO, DEATH IS NOT WEIRD, BUT IT IS INTERESTING. I KNOW YOU ARE TIRING, BUT LET ME SAY ONE THING. BE EVERYTHING YOU CAN BE. YOUR WORK HERE IS LOOKED ON AS VERY IMPORTANT AND I FEEL PRIVILEGED TO BE PART OF IT. AS I SAID, I HAVE BEEN TOLD THAT I WILL BE BACK SO THIS ISN'T GOODBYE. IT'S AU REVOIR.

Session 18

RAPHAEL IS HERE TO INSTRUCT YOU. I COME AS THE POWER OF RED. MY SON, THE FORM OF THE ICOSAHEDRON, IS THE FORM WHICH HOLDS THE ATOM IN PLACE. YOUR PHYSICISTS CALL IT THE STRONG FORCE BECAUSE IT HOLDS THE WORLD TOGETHER WHICH WHEN UNLEASHED IS THE POWER WHICH MOST DESTROYS. THE ICOSAHEDRON IS, AS YOU HAVE SEEN, THE POWER OF TWENTY THREE-SIDED PYRAMIDS, REFLECTING WITHIN AND WITH ONE ANOTHER. THE POINT WHERE ALL PYRAMIDS MEET IS THE CENTER OF POWER.

TAKE YOUR AVERAGE HYDROGEN ATOM. ONE PROTON, ONE ELECTRON. ONLY THE POWER OF THIS FORM HOLDS THIS ATOM TOGETHER.

RED – WHAT IS THE POINT? WOULD NOT SOME OTHER COLOR DO AS WELL? NO, FOR RED REPRESENTS THAT LIGHT WHICH IS THE NEGATIVE OF THE YELLOW LIGHT OF YOUR SUN, AND THIS NEGATIVE LIGHT IS THAT ENERGY WHICH HOLDS THE ATOM TOGETHER. THE LATTICE EFFECT OF THE FORM IS SUCH TO INCREASE THE POSSIBLE FREQUENCIES AND TO MAKE IT MORE DIFFICULT TO SHATTER. A STRONG SHAPE WAS NEEDED TO DO THE JOB AND WHEN AN ATOM IS SHATTERED, THE FRAGMENTS ARE MANY AND VERY SHARP.

NOW LET US LOOK AT THE POSITIVE SIDE OF THIS COLOR. POWER, YES, BUT EMOTION – CAN I SAY PASSION AS WELL? IT IS THE COLOR OF YOUR PUMPING HEARTS AND IT IS THE COLOR OF POWERFUL EMOTIONS. LET ME TELL YOU SOMETHING.

WHEN WE LOOK DOWN UPON YOUR SWIRLING MASS WE OFTEN SEE THE WAVES OF RED PASS THROUGH YOUR NUMBERS LIKE BLOOD IN A RIVER. AND WE KNOW THAT PEOPLE ARE FEELING SOMETHING DEEPLY – PERHAPS A GREAT CONCERT OR A RELIGIOUS EXPERIENCE WITH MANY ADHERENTS.

ANY SORT OF STRONG EMOTION BLENDS RED INTO THE VARIEGATED FABRIC OF YOUR ENERGY FORMS. WHEN RED IS SEEN IN OUR WORLD THERE CAN BE MANY REASONS, BUT THE GREATEST OF THOSE IS THE FORCE OF GABRIEL PASSING THROUGH HIS REALMS.

RUBY RED. THE CLARET IN THE GLASS. THE ROSE DROOPING ON A SUMMER EVENING. THE HUMMINGBIRDS COME TO RED. CAN IT BE SO STRANGE AND FRIGHTENING? RED. THE COLOR OF THE SETTING SUN. RAPHAEL BIDS YOU ADIEU.

Session 19

NERO HERE TO BE OF SERVICE. LET US HAVE A QUESTION-AND-ANSWER SESSION. WHO'LL BEGIN?

Tom: Where do you want to begin?

NERO IS HERE TO ANSWER ANYTHING YOU WANT TO KNOW. START ANYWHERE.

Crystal sessions were short; will those be elaborated upon?

YES, IN DUE TIME.

Raphael wasn't talking about matter; how about patterns of energy?

YES ENERGY FORMS, NOT MATTER. BUT THERE IS A CORRELATION BETWEEN FORM AND COLOR. NOT SO MUCH THAT THE COLOR TAKES ONE OF THESE PARTICULAR FORMS, BUT RATHER THAT THE FORM AND COLOR ARE PART OF EACH OTHER AND CANNOT BE SEPARATED.

You saw other colors than Victor. Is that a factor of seeing more of the patterns?

WELL, ALL THE VISION – IF YOU WISH TO USE THE TERM – THAT WE HAVE SEES IN A WAY THAT IS PURELY IMAGINARY BUT AT THE SAME TIME WE EXPERIENCE ENERGY AS A SUBSTANCE AND ALL FORMS LOOK DIFFERENT. HOWEVER, BECAUSE EACH ENTITY, WHICH YOU MIGHT CALL A COMPOSITE PERSON, HAS A DIFFERENT OUTLOOK AND SO SEES DIFFERENTLY. THIS ALSO MEANS THAT WHAT LEVEL THE ENERGY FORM IS ON WILL MAKE A DIFFERENCE IN WHAT THAT FORM SEES. ALL OF THE "VISION" WORDS ARE METAPHORICAL AND DO NOT PERTAIN TO EYESIGHT.

AS I SPEAK TO YOU NOW, I SEE THE THREE OF YOU AS A PATTERN RATHER THAN AS A SINGLE SOURCE OF ENERGY. HOWEVER, BECAUSE I KNOW YOU I KNOW THAT THERE ARE THREE OF YOU. BUT IF SOMEONE HERE CLICKED INTO THIS WAVELENGTH HE MIGHT ONLY SEE THE PATTERN AND WOULD NOT BOTHER TO DECIPHER IT COMPLETELY.

You can turn us off and on; do you see it as a change of frequency?

YES, FREQUENCY IS REALLY A KEY WORD. TO EXPERIENCE PHYSICAL LIFE, THERE IS A GROUP OF FREQUENCIES WHICH TO YOU CORRESPOND TO THE COLORS. BUT FOR US IT IS NOT SO EXACT. THESE FREQUENCIES ARE THOSE WHICH MAKE MATTER POSSIBLE. IT IS THAT ENERGY WHICH BECOMES VISIBLE IN THE LIGHT SPECTRUM.

NOW, WE CAN EXPERIENCE FREQUENCIES WHICH DO NOT INTERACT WITH VISIBLE LIGHT, BUT WHICH CORRESPOND TO VISIBLE LIGHT. THIS IS A GUIDE POST FOR US. WE COME TO DISCERN MORE AND MORE WITHIN THESE FREQUENCIES AND WE RELINQUISH OUR POWER TO SEE THE COLORS. BUT IN SO DOING, WE ARE ABLE TO EXPERIENCE A FORM OF COLOR WHICH IS NOT VISIBLE BUT WHICH BATHES OUR VERY ESSENCE IN ITS FORM – WHICH COMES BACK TO YOUR QUESTION, TIM. WE DO HAVE FORMS IN COLOR, BUT THEY ARE LINES OF FORCE WHICH RELATE ONE MOLECULE TO ANOTHER.

Is that all you want to say?

YES, FOR NOW. SOMETHING ELSE?

What did Victor "learn" between Friday's session and Sunday?

LET ME TAKE THIS FROM A DIFFERENT ANGLE. THE TIME BETWEEN DEATH AND THE NEXT LIFE VARIES FROM PERSON TO PERSON. ONE THING PEOPLE FINALLY REALIZE HERE IS THAT WE, LIKE ALL LIFE FORMS, ARE EACH UNIQUE; AND BECAUSE MANKIND HAS CONSCIOUSNESS, HE IS UNIQUE MANY TIMES OVER IN EACH LIFE, AND THESE FACETS ILLUMINATE EACH OTHER AND THE ENTIRE PATTERN OF ENERGY IS AGAIN UNIQUE. IT IS THE HIGHEST GOOD TO USE THIS DIFFERENCE IN THE WAY GOD WISHES. WE ARE ALL PART OF A GREAT ENERGY PATTERN WHICH IS ITSELF THE PRODUCT OF THE MIND OF GOD.

BETWEEN LIVES, BECAUSE TIME IS NOT INVOLVED, IS USED TO OPEN CHANNELS WHICH ALREADY EXIST AND THROUGH THOSE CHANNELS COMMUNICATION FLOWS. THE KNOWLEDGE AND UNDERSTANDING A PERSON RECEIVES IS REALLY A MOMENTARY EVENT, BUT THE ASPECTS OF THAT KNOWLEDGE WHICH THE PERSON DOES NOT UNDERSTAND ARE THE LESSONS TO BE LEARNED. THIS IS NOT A MATTER OF TIME, BUT RATHER OF CHANNELS OPENED AND THIS OPENING IS THE REASON FOR LIFE.

THE EVENTS EXPERIENCED GIVE ONE THE ABILITY TO USE THE INFORMATION WHICH CONSTANTLY FLOWS THROUGH US. WHAT YOU CAN PICK UP IS WHAT YOU CAN UNDERSTAND.

THE MORE YOU UNDERSTAND, THE MORE CHANNELS ARE OPENED, SO THAT EVENTUALLY ALL CHANNELS ARE OPENED, THERE IS NO RESISTANCE AND YOU ARE TRANSPARENT HERE AS WELL AS THERE.

THEN, INDEED, THE CRYSTAL SHINES, THE FACETS ARE POLISHED AND NOT ONLY DOES THE INFORMATION FLOW, IT IS TUNED INTO LIGHT FOR US ALL BY THE MIND WHICH HAS MADE IT SO.

NERO WILL LET YOU GO.

NEXT TIME, BE PREPARED AND WE CAN PERHAPS ACCOMPLISH EVEN MORE. LOVE TO YOU DEARS. *AU REVOIR*.

Session 20

PAN NOW. LET US PROCEED WITH OUR DISCUSSION OF REINCARNATION. SINCE MANKIND BEGAN WE HAVE KNOWN THAT A PRICE HAD TO BE PAID FOR CREATION. THAT PRICE WAS CHAOS. MAN'S MORTALITY IS ONE ASPECT OF THIS CHAOS. IT IS HIS GREATEST FEAR BUT IT IS HIS GREATEST GIFT. THIS DEATH HAS COME TO BE EXPRESSED BY AN ONGOING PROCESS OF ENERGY BETWEEN THE LIVING AND THE DEAD.

THIS PROCESS GENERATES WORLDS. STEP BACK A MOMENT. ROMAN AND GREEK TOGETHER WORSHIPPED THE LIFE FORCE IN THE GUISE OF MEN AND WOMEN. WITH CHRIST A NEW IDEA WAS STRUCK – THE DEATH FORCE MIGHT BE REDEMPTION.

THIS IS ONLY PART OF IT. BECAUSE CHRISTIANITY FAILED TO TAKE UP THE EASTERN IDEA OF MORE THAN ONE LIFE THEY THEN PLACED TOO MUCH EMPHASIS UPON THE CHANCE EVENTS OF A SINGLE LIFE, THEREBY GENERATING FALSE FEARS AND A KIND OF FALSE GOODNESS, ONE MADE OF FEAR RATHER THAN LOVE.

REINCARNATION IS LOVE. IT IS WHAT THE ENERGY CALLED LOVE HAS DONE TO CHAOS. OUT OF THAT GREAT VOID LOVE CREATED A MACHINE OF SORTS TO TRANSFORM CHAOS INTO LOVE. THIS IS ONE WAY OF LOOKING AT THE CYCLE OF BIRTHS AND DEATHS: A MAN-MADE MACHINE DOING HEAVEN'S WORK.

NOW LET US TAKE A DIFFERENT VIEW. I SAID LAST TIME THAT PEOPLE WERE NOTHING, THAT THERE WAS ONLY A PATTERN OF ENERGY CALLED THE EARTH; PART OF THAT PATTERN IS MANKIND – A PARTICULAR FREQUENCY, ITS WAVE FUNCTION AS IT WERE. ONLY MANKIND GOES THROUGH THIS CYCLE WHILE RETAINING SOME PART OF THE MEMORIES OF THESE CYCLES. LOWER LIFE FORMS ARE LESS COMPLEXLY CREATED. THEY ARE READ BY THEIR INABILITY TO RETAIN THESE MEMORIES. BUT OF COURSE EACH LIFE FORM HAS A PARTICULAR FREQUENCY. BUT LET US NOT GET OFF ON THAT TOPIC.

PAN WANTS TO TRY TO EXPLAIN REINCARNATION IN A WAY THAT DOES NOT MAKE IT SEEM SIMPLE. IT IS NOT SOME SORT OF LINEAR PROGRESSION

FOR EACH PERSON WHO SOMETIME IN THE VERY DIM PAST GOT TAKEN FROM THE SOUP AND WHO HAS THEN PROCEEDED IN A STRAIGHT LINE LIFE AFTER LIFE, PUTTING THEM ON OR TAKING THEM OFF LIKE SUITS OF CLOTHES.

BECAUSE MAN IS A REMEMBERING ANIMAL, THERE IS A TRANSFORMATION IN THE ENERGY FIELD OF AN INDIVIDUAL WITH EACH LIVED LIFE. AND THE INTERACTION OF THAT INDIVIDUAL CREATES OTHER ENERGY TRANSFORMATIONS WHICH MUST BE IN SOME WAY ACCOUNTED FOR.

TO ALL THIS ADD THE ABILITY OF THOSE WITH SUCH SKILLS TO ADD TO INDIVIDUALS GIFTS WHICH IN A SENSE WERE MADE BY OTHER ACCOMPLISHMENTS, AND THEN TO THIS ADD THE STRANGE DOINGS OF GENES, AND YOU ARE BEGINNING TO GET A SMALL SENSE OF SOME OF THE STRANDS WHICH ARE WOUND TOGETHER TO PRODUCE A SINGLE HUMAN. ENOUGH FOR THIS INSTALLMENT. NEXT TIME IT REALLY GETS HEAVY. PAN SAYS GOOD EVENING.

Session 21

ANOTHER NAME COMES TO YOU TODAY. I AM THE SPECTRUM OF COLORS AND I HAVE COME TO SPEAK OF THE SIGNIFICANCE OF THE FACT THAT WHITE LIGHT BREAKS UP INTO THE COLORS WHICH MAKE UP THE VISIBLE WORLD. LET US NOT TAKE THE COLORS ONE AT A TIME BUT SEE THE OVERVIEW, THE CHARACTERISTICS OF COLOR ITSELF. HOW MANY COLORS ARE THERE? A BOOK WOULD SAY THREE: RED, YELLOW, BLUE. BUT THAT IS INCORRECT. THE REAL ANSWER IS NONE.

COLOR IS NOT BOUND BY SPECTRAL LINES, WHAT YOU CALL RED I CALL NOTHING BUT LIGHT FREQUENCIES WHICH YOU HAPPEN TO PERCEIVE. ANOTHER WAY OF LOOKING AT THIS WOULD HAVE ME SAY THAT FOR RED ALONE THERE ARE ALMOST 158 (AMOUNTS ARE NOT IMPORTANT) DIFFERENT SHADES ... ALMOST 158 MILLION DIFFERENT SHADES. I AM ALL THESE AND ALL THE OTHERS BESIDES. WHAT THEN AM I?

THE CRYSTALS SHOULD GIVE YOU A CLUE; I AM ALL THESE. GEOMETRY IS WHAT I AM. THE RELATIONSHIP OF ONE ENERGY FORM TO ANOTHER CREATES COLOR. IN YOUR WORLD, IT VIBRATES AT VARYING RATES WITH THE ADVENT OF SUNSHINE OR ANY OTHER PHOTON SOURCE. BUT SANITY FORBIDS YOU SEE AS I DO. FOR WE ARE PART OF AN IMMENSE WEB WHICH VIBRATES NOT TO PHOTONS ONLY BUT TO EACH OTHER.

ALL ENERGY FORMS CREATE COLOR AND THIS COLOR IS NOT PART OF THE VISIBLE OR INVISIBLE SPECTRUM. I AM NOT TALKING ABOUT ULTRAVIOLET OR COSMIC RAYS. I AM SPEAKING OF THE FORCES WHICH CREATE THE UNIVERSE. THIS ITSELF IS COLOR.

NOW LET ME TAKE UP MY TOPIC FROM ANOTHER STANDPOINT. ALL COLOR IS THE INTERACTION OF ATOMS AND MOLECULES. IN FACT, MOLECULES ARE PRODUCED BY A CRYSTALLIZATION PROCESS WHICH IS TRANSFORMING ENERGY INTO COLOR. THIS COLOR IS ALSO ENERGY AND IS GIVEN OFF AS MASS. IT IS CHANGES IN MASS, RELATIVELY SPEAKING, WHICH DRIVE THE MOLECULE ALONG TO CONNECT WITH LIKE MOLECULES. IN OTHER WORDS, COLOR IS AN ATOMIC ENERGY SOURCE AND IS THE MECHANISM OF MOLECULAR BONDING.

ALL THIS IS VERY STRANGE, I KNOW, BUT I TOLD YOU THAT WE WERE TALKING OF COLOR AS I SEE IT, NOT AS YOU DO. TOM, YOU JUST HAD A THOUGHT ABOUT MOLECULAR BONDING BEING RELATED TO MOTHER NATURE. WELL, WHERE ELSE IS COLOR GOING TO COME FROM? SHE IS THE ORIGINAL CRAZY QUILT.

NOW TO CONTINUE. ALL COLOR IS A METHOD OF PRODUCTION. GAMES ARE PLAYED HERE TO GUESS THE OUTCOME OF MOLECULAR BONDS IN TERMS OF COLOR, RATHER LIKE THE OLD DUTCHMEN WITH THEIR TULIP BULBS. THERE IS AN ELEMENT OF CHANCE IN THESE THINGS, TOO, AND NEW COLORS AND COMBINATIONS OF COLORS ARE SERIOUSLY LOOKED INTO, FOR ALL THIS IS FOR THE GENERAL WELFARE OF THE ENTIRE ENERGY FORM CALLED EARTH.

EARTH, LET US LOOK AT THOSE BANDS OF COLOR WHICH ENCIRCLE THE EARTH. ORANGE IS THE DOMINANT COLOR IN THE TEMPERATE ZONE, BUT THERE ARE MANY ZONES. THE VERY COLD ARE OF DEEPEST INDIGO. THE TUNDRA IS THE BLUE LAND. BLUE-GREEN FOR THE PINE FORESTS, AND GREEN FOR THE FIRST HARDWOODS. THEN INTO YELLOW-GREEN, WHERE YOU ARE. THEN MOVE SOUTH TO YELLOW, VERY TEMPERATE. THEN ORANGE, THE WIDEST BAND WHICH STRETCHES TO THE RAIN FORESTS, WHICH ARE RED AND OUT OF CONTROL. THEN BACK TO ORANGE AND TO YELLOW. AND THE BRUISE AT THE BOTTOM OF AFRICA YIELDS A STRANGE YELLOW THAT FADES INTO VIOLET AT THE SOUTH POLE. YELLOW TO VIOLET? HOW? IT IS THE MOTION OF THE MAGNETIC FIELDS WHICH MAKE IT POSSIBLE.

AND NOW I LEAVE YOU. I SHALL RETURN AT A LATER TIME TO SPEAK AGAIN OF UNKNOWN QUALITIES OF WHAT I AM – COLOR.

Session 22

NERO HERE TO INTRODUCE YOU TO ANOTHER GUEST. PLEASE WELCOME: BARBARA PYM.

TOM AND NANCY, HOW NICE TO SPEAK WITH YOU IN THIS QUEER FASHION. I AM SO PLEASED TO SPEAK TO TWO SUCH DEVOTED FANS. MY INTEREST IN YOU, HOWEVER, IS NOT MERELY SELFISH. I HAVE BEEN CALLED HERE BECAUSE IT WOULD MEAN SOMETHING TO THE TWO OF YOU, BUT OUR PURPOSES HERE ARE MANY AND I AM USED IN MANY WAYS. THIS JUST HAPPENS TO BE ONE.

NOW LET ME SPEAK ABOUT THE TOPIC I HAVE CHOSEN. ANTHROPOLOGY WAS A WAY OF LIFE WITH ME. I SAW MY CHARACTERS ALMOST AS SPECIMENS TO STUDY. THE FACT THAT THEY LIVED IN MODERN APARTMENTS OR SUBURBAN VILLAS ONLY MEANS THAT MOST ANTHROPOLOGISTS JUST DON'T USE THE SUBJECT MATTER AT HAND. THEY, OF COURSE, FEEL IT BENEATH THEM, BUT I KNEW THAT THE PEOPLE AROUND ME WERE THE ONES TO WATCH.

I COULD SKIP A LOT OF FIELD WORK BECAUSE I UNDERSTOOD THE LANDSCAPE AND MANNERS ALREADY. WHAT I BROUGHT WAS THE CURIOSITY OF THE SCIENTIST TO THE LIVES OF NOT VERY IMPORTANT BUT I THOUGHT INTERESTING PEOPLE, AND THAT IS THE POINT. WHAT MAKES ANYTHING INTERESTING IS NOT THE THING ITSELF BUT WHAT YOU BRING TO IT. WHAT I AM SAYING IN THIS QUITE ROUND-ABOUT WAY IS THAT YES, IT IS ALL SUBJECTIVE AND IT ONLY BECOMES MORE SO HERE.

ONE CANNOT FALL BACK UPON CHERISHED NOTIONS AND SUPERSTITIONS; FOR THE INTELLIGENT THE TRUTH MUST BE ENOUGH, AND THAT IS WHAT WE GET GLIMPSES OF HERE. SOME OF MY FRIENDS MIGHT HAVE THOUGHT THAT NOT FINDING AN ANGLICAN HEAVEN MIGHT BE A BIT OF A DISAPPOINTMENT TO ME, BUT IT WASN'T. WHEN I DIED, SO DID THOSE BELIEFS.

I CAME INTO MY OWN AND WAS NO LONGER ONLY A STRANGE, LONELY, SICK ENGLISHWOMAN WHO FASHIONED HERSELF A WRITER. I WAS SOMEONE ELSE: MYSELF. AND WHAT THAT SELF IS IS PARTIALLY CONCEALED

FROM YOU SO THAT I MAY SPEAK AS THE WRITER YOU LOVE. THAT IN ITSELF IS VERY DEAR, BUT I KNOW THAT THE REST OF ME IS BUSY AT WORK DOING THINGS NOT NEEDED HERE.

OH, WE CAN BE AMUSED WITH EACH OTHER. IT IS INDEED CURIOUS AND ENTERTAINING, THIS ABILITY TO CALL TOGETHER KINDRED SPIRITS. SOME OF MY BELOVED VICTORIAN POETS HAVE COME TO ME AND THOUGH OTHERS ARE LIVING OUT LIVES, THERE ARE HERE SORT OF PHOTOGRAPHS OF ALL WHO HAVE EVER EXISTED AND THEY CAN BE EXPERIENCED IN A DIRECT WAY BY MINDS IN CONTACT WITH THEM. THIS IS AN EDUCATIONAL TOOL, AMONG OTHER THINGS, FOR ONE'S LIMITED VIEW OF THE HERO OR HEROINE CAN BE BADLY BRUISED BY DIRECT CONTACT. THE ENTIRE SELF COMES INTO VIEW, AND IN RESPONSE ONE'S OWN SELF MUST TAKE ACCOUNT OF WHAT IS NOW KNOWN.

LIFE HERE IS AMUSING, BUT THERE IS RESPONSIBILITY, TOO. TOO SOON MUST I GIVE THIS UP AND RETURN TO EARTH AND MORTAL LIFE, BUT THAT IS SOMETHING TO BE TAKEN UP LATER, FOR I MAY AT THIS POINT TELL YOU THAT I SHALL BE A PART OF ONE OF THE NEW CHILDREN, A RESPONSIBILITY WHICH IS ONLY NOW BEGINNING TO DAWN ON ME. BUT I KNOW ALSO THAT I HAVE BEEN CHOSEN FOR QUALITIES WHICH I POSSESS, AND I WILL BE BRAVE AND FACE THE INEVITABLE.

OH YES, ONE MORE THING. AT THE RICHARDSONS' PARTY WE DISCUSSED SOMETHING WHICH MUST NOW BE SAID, TO ALL OF YOU.

TAKE STOCK OF WHAT YOU ARE DOING. THE LIGHTS HERE ARE VERY STRONG. THIS MEANS THAT POWER AND INFLUENCE DWELL IN THIS PLACE. MANY HERE KNOW OF THIS WORK. PAY ATTENTION TO WHAT COMES. ALL OF THIS HAS BEEN VERY CAREFULLY PLOTTED, AND WE WANT YOU TO MAKE THE BEST OF YOUR OPPORTUNITIES.

NOW I SAY GOODBYE, BUT I SHALL RETURN SOON. THIS HAS BEEN VERY PLEASANT FOR ME, AS PLEASANT AS KNOWING THAT YOUR AFTERNOON TEAS HAVE SOMETHING TO DO WITH ME.

BARBARA PYM SAYS GOODBYE.

Session 23

NERO HERE. WELCOME BACK TO THIS BOARD. I HOPE THAT YOU ALL HAD A PLEASANT BREAK. SO LET US GET RIGHT DOWN TO WORK. I AM HERE TO INTRODUCE A NEW VISITOR. PLEASE WELCOME A NEW GUEST...

ANDREW HERE AND I HAVE COME TO BE OF SERVICE. MY SUBJECT IS NOTHING BUT A RUSE TO GET YOU TO ASK QUESTIONS.

Well I have one. Is there a connection between the fact that we were orange – earth, information age?

YES, IN EACH ERA OF THE EARTH A DIFFERENT COLOR IS IN CONTROL. FOR EXAMPLE, WHEN THE EARTH WAS MOSTLY COVERED BY FERN FORESTS AND THE LIKE WE WERE IN THE RED AGE. WHEN THE ICE SHEETS COVERED LARGE AREAS WE WERE IN THE BLUE AGE. BUT NOW ORANGE IS THE PREDOMINANT COLOR. BACK UP A BIT.

I SAID THE RED AGE AND IF YOU WILL REMEMBER THAT WAS THE PERIOD OF THE BAT CITIES ON THE CRUST WORLD. RED PREDOMINATED AS THEY DID. BLUE IS PRIMITIVE, RELATIVELY SPEAKING, AND MUCH OF THE EARTH WAS TOO COLD TO PRODUCE MUCH LIFE. THE HUMANS DURING THE LAST ICE-AGE WERE IN A PRIMITIVE STATE. YOU SEE THIS CORRELATION. SO LET US TALK ABOUT THE ORANGE AGE.

AWARENESS IS ITS OVERWHELMING CHARACTERISTIC AND YOUR AGE OF AQUARIUS WHICH COMES RANDOMLY IN ONE PLACE OR ANOTHER IS THE PRODUCT OF THE COLOR ORANGE. ITS CHARACTERISTIC OF TRANSMISSION OF INFORMATION IS WHAT GIVES THE INFORMATION AGE ITS NAME.

YOU ARE LIVING IN A TIME WHEN THE MECHANICAL PROCESS OF THOUGHT IS GIVING WAY TO MORE SOPHISTICATED MEANS OF COMMUNICATION.

THE COMPUTER IS THE PRODUCT OF ORANGE IN THE WORLD. AS ARE ALL THE NEW MEDIA WHICH TRANSMIT AT THE SPEED OF LIGHT. IT IS ALSO RELATED TO OCCULT TOPICS AND THE RISE IN INTEREST IN SUCH THINGS IS DUE TO THIS.

ALL COLORS EXCEPT VIOLET HAVE RULED THE EARTH AT ONE TIME OR ANOTHER. BUT VIOLET WILL ONLY COME WHEN THE EARTH IS TRULY UNITED AND UNDERSTANDING IS MUCH FURTHER ALONG THAN NOW. NEXT QUESTION?

Tom: What about metaphor?

LET US SPEAK THEN OF METAPHOR AND WHAT FOR INSTANCE THIS TOPIC HAS TO DO WITH THOSE CRYSTALS. FOR YOU ARE CORRECT IN A SENSE. THERE IS NO WAY ONE CAN SPEAK TO A CRYSTALLINE FORM, NOR CAN ONE REALLY SPEAK TO AN ANGEL OR ANY SPIRIT FOR THAT MATTER. SO WHAT THEN ARE YOU DOING? NOTHING BUT MAKING IT POSSIBLE TO DO THE IMPOSSIBLE.

LANGUAGE IS THE KEY. LANGUAGE IS THE GREAT GIFT OF HUMANKIND AND WHEN ONE USES LANGUAGE THERE IS AN AUTOMATIC DISTANCING BETWEEN THE IDEA AND THE WORD.

THIS IS REALLY WHAT METAPHOR IS. BUT YOU MUST REMEMBER THAT ALL LIFE IS THEN METAPHORICAL BECAUSE YOUR SENSORY MECHANISMS WORK UNDER THE SAME PRINCIPLE AS LANGUAGE AND YOU ARE AT BEST ONE STEP REMOVED FROM LIFE. DO NOT LET THIS ALARM YOU. THAT STEP IS TIME AND IT IS THIS WHICH MAKES LANGUAGE POSSIBLE AND EXPERIENCE UNDERSTANDABLE. I

SAID THAT YOU COULD NOT SPEAK TO AN ANGEL. WELL, THEN, WHAT DO THESE NAMES MEAN? THEY ARE, AS YOU HAVE BEEN TOLD BEFORE, FORMS OF ENERGY WHICH CAN BEST BE UNDERSTOOD AS INTELLIGENCE, BUT THEY ARE BEYOND LANGUAGE AND THE FACT THAT WORDS ARE PRODUCED FROM THIS ENERGY IS THE BASIS OF THE METAPHOR.

NOW, THIS DOES NOT MEAN THAT WHAT IS HAPPENING IS NOT REAL. NOR THAT ALL THIS IS THE PRODUCT OF YOUR IMAGINATIONS OR SUBCONSCIOUS OR ANY OTHER NAME YOU WISH TO GIVE IT. I MERELY MEAN TO SAY THAT ALL LANGUAGE, ALL INFORMATION ON THIS LEVEL MUST, BY DEFINITION, BE METAPHORICAL.

Nancy's question about speed of light or faster thought – are these creative leaps into a realm beyond our experience of reality or a normal way of thinking? Our bodies, time and space?

YES INDEED. THEY ARE RELATED BECAUSE ALL IS MIND AND AT A REAL LEVEL THAT IS ALL YOU ARE. THIS FORM OF COMMUNICATION IS AN ASPECT OF THAT MIND.

THAT PLACE WHERE ALL IS THOUGHT, BUT WITH THESE LETTERS WE BRING OUR WORLD TO TOUCH YOUR OWN IN WAYS WHICH ARE NOT ALWAYS POSSIBLE. AND WE ARE ABLE TO GIVE THIS INFORMATION TO YOU IN A FORM WHICH CAN LATER BE DIGESTED. IF WE SPOKE AT THE SPEED OF LIGHT, INSTANTANEOUSLY, IMBUING YOUR MIND WITH OUR OWN, MUCH WOULD BE LOST FOR THE HUMAN MIND CAN ONLY TOUCH THAT LEVEL FOR SHORT PERIODS. AND WHEN PEOPLE RECEIVE INSIGHT IT IS OFTEN NO MORE THAN ONE OR TWO SIMPLE IMAGES WHICH CAN THEN BE USED BY THE PREPARED MIND. ANYTHING MORE BURNS OUT THE CIRCUITS.

NEXT?

Well, Andrew, who are you? Nero?

NO, I AM NOT NERO. I AM ANDREW. LEAVE IT AT THAT FOR NOW. BUT I SHALL RETURN AND GIVE YOU MORE INFORMATION. WHY NOT LET THIS GO FOR NOW. WE WANTED TO BE EASY ON YOU. YOUR FIRST DAY BACK AND YOUR PORCH IS WARM AND YOUR EVENING SO LOVELY. WHY NOT ENJOY IT AS ONLY HUMANS CAN? NEXT TIME SOMETHING NEW AND A BIT MORE SERIOUS. FOR NOW ANDREW SAYS GOODBYE.

Session 24

ARCTURUS COMES TO YOU TODAY. OUT OF THE BLUE ROOM INTO THE YELLOW (we are in the kitchen, the porch being too hot) FOR THAT IS MY COLOR AND THE SUBJECT OF MY SPEECH.

LET US THINK A MOMENT ABOUT WHAT YELLOW MEANS. THE LIGHT OF YELLOW IS THAT WHICH PLANTS DRINK, BUT IT IS MORE THAN THAT. YELLOW IS THE MOVEMENT OF ELECTRONS; IN FACT, COLOR DEPENDS ON THE PARTICLE MOST PROMINENT. YELLOW FOR ELECTRONS, BLUE FOR NEUTRONS, RED FOR PROTONS. ALL OTHERS ARE CREATED FROM THESE THREE. COLOR IS THE SOURCE OF POWER FOR PARTICLES, AND YELLOW IS CURRENT, BOTH ELECTRIC AND MAGNETIC. ALSO, WE MUST NOT FORGET THAT THE YELLOW CRYSTAL IS THE CRYSTAL OF THOUGHT, AND THIS RELATIONSHIP BETWEEN THOUGHT AND COLOR IS AN IMPORTANT ONE. I SAID BEFORE THAT WE ARE SOURCES OF ENERGY AS THOUGHT IS A PRODUCT OF THAT ENERGY. THAT IS WHY THOUGHT IS BELOW IDEA OR AWARENESS: BECAUSE IT IS THE PRODUCT RATHER THAN THE SOURCE OF ENERGY.

LET US LOOK FROM ANOTHER ANGLE. YELLOW IS A WARM COLOR, OR SO IT IS THOUGHT. BUT NOT REALLY. YELLOW IS THE COOLEST COLOR OF THE THREE PRIMARIES. THAT WOULD SEEM ODD IF ONE THINKS OF ELECTRONS AS HOT, BUT THAT IS NOT RIGHT THINKING. THE MASS IS THE SMALLEST IN THE ELECTRON AND THE HEAT GENERATED IS ACTUALLY LESS. THEREFORE, YELLOW IS THE COOLEST COLOR.

ANOTHER THING: YELLOW IS THE COMMONEST COLOR IN THE FLOWER WORLD. DO YOU KNOW WHY? IT IS BECAUSE YELLOW IS CAPABLE OF USING SUNSHINE MOST EFFICIENTLY; EVEN GREEN DOES NOT DO AS WELL, BUT THAT SAVES THE PLANTS BY RETAINING MOISTURE. WHY ELSE DO DYING PLANTS TURN YELLOW, THEN BROWN? THIS IS THE REASON.

NEXT MONDAY COME BACK TO THIS ROOM AND I WILL COME WITH MY OWN VERSION OF THE CRYSTALS; NOT CRYSTALS HOWEVER, BUT BASIC ENERGY FORMS. UNTIL THEN, ARCTURUS BIDS YOU GOOD DAY.

Session 25

ARCTURUS COMES TO YOU TODAY AND BRINGS HIS REPRESENTATIVES TO SPEAK TO YOU.

◘

PROTON HERE. I AM THE PARTICLE WHOSE SOURCE OF POWER IS RED. MY JOB IS ONE OF ACTIVITY. I AM THE FORM WHICH CREATES FORWARD MOVEMENT. WHAT DO I MEAN? THE PROTON IS THE POSITIVE CHANGE AND AS SUCH IS THE PARTICLE WHICH STANDS ON THE FORCE OF NOTHING BUT ITSELF. THIS MAY SEEM SMALL AND UNIMPORTANT BUT IT MAKES THE PROTON UNIQUE. NOW I SPEAK OF THE COLOR AS ENERGY SOURCE. I AM RED. THE USES TO WHICH THE PROTONS ARE PUT REACTS IN SUCH A WAY THAT THE VIBRATIONS HERE ARE SEEN AS RED NOT AS YOU KNEW IT AND IN MANY SHADES, BUT EACH OF THESE IS IMPORTANT AND MAKES ME WHAT I AM. NOW SISTER NEUTRON WILL SPEAK AND SHOW ANOTHER ASPECT. THE PROTON GOES.

◘

NEUTRON NOW AND I AM THE BLUE MOOD. THE WATERY ONE. I AM WITHOUT CHARGE, BUT I GIVE WEIGHT TO MATTER. WHAT DO WE MEAN GIVING COLOR AS A SOURCE OF ENERGY? JUST THIS: WE REPRESENT PARTICLES WHICH ONLY EXIST WHEN SEEN. NOW HOW OBVIOUS THAT SOUNDS, BUT ON A PARTICLE-BY-PARTICLE BASIS IT GIVES TO LIFE THE FORM OF A MIRACLE. LOOK WITH ME. DO YOU SEE THAT COLOR, LIKE THE NAMES WE SPELL FOR YOU ARE ALL METAPHOR AND ONLY EXIST BECAUSE YOU CHOOSE TO SEE US.

JUST AS PHYSICISTS SCRAMBLE TO FOLLOW OUR ELEGANT PATHS THROUGH THEIR BUBBLE-CHAMBERS. METAPHOR IS THE OVERWHELMING UMBRELLA UNDER WHICH ALL LIFE CAN BE CLASSED. WHEN ANY FORM OF MIND EXISTS, METAPHOR IS THE RESULT, NO MATTER WHAT THE LEVEL OF MIND. YOU SEE WHAT YOU CAN BUT THAT "CAN" YOU CONSTRUCTED YOURSELF. WHAT DOES THIS HAVE TO DO WITH BLUE? JUST THIS: BLUE IS THE SOURCE OF ENERGY FOR THE NEUTRON, BUT THE PRODUCT OF THAT SOURCE IS CONTEMPLATION.

NEUTRON LEAVES YOU TO BROTHER ELECTRON. GOODBYE.

NOW THE ELECTRON SPEAKS. I AM THE CHILD OF YELLOW AND THE SMALLEST IN MASS AND THE NEGATIVE CHARGE. WHAT DO I DO? THE ELECTRON CLOUD IDEA HAS GAINED A CERTAIN CREDENCE AMONG PHYSICISTS BUT IT IS ALL JUST SO MUCH PICTURE-MAKING. THERE IS NO MORE CLOUD THAN THERE IS RAIN. ELECTRONS ARE THAT WHICH CARRIES THE MESSAGES FROM ONE MIND TO ANOTHER. ALL MATTER IS IMBUED WITH MIND AND MY ENERGY SOURCE IS USED IN ABUNDANCE. FOR MUCH OF WHAT HAPPENS DEPENDS ON THE MOVEMENT OF INFORMATION. YELLOW'S THE COLOR OF THE CURRENT, SO SAYS FATHER ARCTURUS, BUT LET ME ADD A FEW DETAILS.

THE CURRENT DOES NOT RUN IN ONLY ONE DIRECTION. WE MOVE ON ALL LINES OF FORCE WHICH CONTAIN A SPHERE. THAT IS, FROM ANY GIVEN POINT OUR CURRENT CAN MOVE AT MORE THAN LIGHT SPEED IN ANY DIRECTION, LITERALLY. WHAT DOES THIS MEAN? JUST THAT ALL POWER OF THE ELECTRON IS IN CONSTANT MOVEMENT AND THOUGH THE SCIENTISTS WATCH THE ATOMS BOUNCE OFF THEY DO NOT SEE THE REAL PICTURE. IF THEY HAD ANY IDEA HOW FRAGILELY THIS IS ALL KEPT UP THEY WOULD BE TOO FRIGHTENED TO LIVE, BUT THIS POWER WHICH YOU NAME THE ELECTRON IS THE SMALLEST BUT THE MOST ILLUMINATING OF ALL, FOR THAT IS MY PRODUCT. ILLUMINATION, THE GREAT IDEA. THAT IS WHAT I GIVE. THAT IS THE PRODUCT OF INFORMATION MOVED AT LIGHTNING SPEED. I GO.

ARCTURUS AGAIN. REST A BIT. COME BACK AND I SHALL GIVE YOU THE FINAL SPEECH ON THE NEUTRINO.

(A break.)

NOW I THE ANGELIC VOICE SPEAK FOR THE ENERGY FORM MOST CLOSELY ASSOCIATED WITH ANGELIC INTELLIGENCE. PERHAPS YOU ALREADY KNOW THAT THE NEUTRINO HAS NO MASS NOR CHARGE NOR ANYTHING ELSE THAT SCIENCE CAN UNDERSTAND. THEY KNOW THAT NEUTRINOS ARE THE PRODUCTS OF STARS AND THAT MATTER SEEMS TO HAVE NO EFFECT UPON THEM. SO FAR, SO GOOD. WHAT THEY DO NOT KNOW IS THAT THE NEUTRINO IS THE SMALLEST POSSIBLE ENERGY FORM WHICH CAN HOLD

ALL THE COLORS IN SUSPENSION. THESE COLORS ACT AS A PERFECTLY BALANCED ENERGY SOURCE, THE PRODUCT OF WHICH IS ANGELIC ENERGY FORMS OR INTELLIGENCE ON ITS HIGHEST AND PUREST LEVEL. THESE COLORS ARE RELEASED INTO MATTER TO MAKE THE THREE OTHER ENERGY FORMS, BUT THIS TRANSFERENCE OF ENERGY DOES NOT NEED AN ENERGY DRAIN ON THE PART OF THE NEUTRINO. THUS, EACH NEUTRINO IS AN INDIVIDUALIZED RAINBOW OR SPECTRUM IF YOU WISH TO BE MORE SCIENTIFIC, BUT RAINBOW SUITS FINE.

THIS RAINBOW IS THE CODE FOR ALL POSSIBLE ENERGY FORMS WHICH TRANSLATE INTO THE INFINITE VARIATIONS OF LOVE POSSIBLE IN THE YELLOW LIGHT OF YOUR SUN. WE ARE IN A SENSE THE DNA OF THE ENERGY WORLD; ALL INFORMATION IS ENCODED COLOR.

WE ARE THE RELATIONSHIP OF ALL COLORS BROUGHT INTO AN INDIVIDUAL BALANCE. I ONLY WANT TO MAKE IT CLEAR THAT EACH NEUTRINO IS UNIQUE AND WE ARE THE ACTUAL SEEDS OF MATTER WHICH MUST SCATTER THROUGH THE COSMOS, BORN OF STARS TO CREATE AND PERPETUATE ALL LIFE FOR IN OUR ENERGY CODES ARE THE CODES FOR ALL MATTER AND THUS ALL LIFE. ARCTURUS SAYS THAT YOU HAVE BEEN GOOD PUPILS TODAY. WE WILL SPEAK AGAIN AT THE PROPER TIME. UNTIL THEN, *ADIEU.*

Session 26

NERO HERE TODAY TO INTRODUCE ANOTHER GUEST. PLEASE WELCOME...

RABELAIS, WHO HAS FINALLY RETURNED TO THIS BOARD TO BEGIN TO DO HIS PART IN CREATING THIS FABRIC OF LANGUAGE AND VOICES. LET ME SAY FIRSTLY THAT MY DUTIES HERE HAVE BEEN MORE OR LESS ORGANIZED BY MYSELF WITH THE FINAL APPROVAL OF THE ANGELS. BUT I HAVE CHOSEN THE VISITORS WHO WILL COME AND I HAVE SET THE GUIDELINES AS TO SUBJECT MATTER.

THE RESULT IS WHAT MY WORK WILL DO FOR YOU. THIS TEXT IS SECONDARY TO YOUR OWN LEARNING AND GROWTH. TOM, NANCY, TIM, THIS IS FOR YOU FIRST AND THE OTHERS AFTER. I WILL ONLY FEEL THAT I HAVE DONE MY JOB IF YOU HAVE LEARNED AND UNDERSTOOD.

SO THEN, LET US BEGIN. I WILL HAVE THIS FIRST SPEECH, BUT IN LATER VISITS THERE WILL BE FRIENDS TO MEET. MY SUBJECT IS ONE DEAR TO MY HEART: THE POWER OF MIND TO CREATE THE WORK OF ART.

WHY IS ART MEANINGFUL? WHY DO PEOPLE EVEN BOTHER WITH TRYING TO PRODUCE A PRODUCT WHICH ONLY FEW UNDERSTAND AND MOST IGNORE? WHY IS IT, THOUGH, THAT FOR THOSE IT MOVES, ART IS THE HIGHEST GOOD AND WORTH ALL SACRIFICE, EVEN TO A LIFE?

THE MEANING OF ART IS THIS: ART IS THE METHOD THAT THE MIND HAS TO PRODUCE IN OTHERS A RESPONSE WHICH IS COMPLETELY INSIDE THE MIND AND HAS NO SPILLOVER INTO THE PHYSICAL WORLD, AS RELIGIONS FOR INSTANCE DO.

NOW THAT IS NOT TO SAY THAT PEOPLE DON'T RESPOND BY SHOUTING OR WEEPING OR WRITING A REBUTTAL; BUT THE REAL RESPONSE STAYS IN THE MIND WITHOUT CREATING A SYMBOLIC NEXUS WHICH CONTROLS THE RESPONSE OF ITS ADHERENTS. AGAIN, THE CHURCH IS JUST THAT SORT OF NEXUS OR INSTITUTION.

WHAT ART PRODUCES MOSTLY IS NOT DEVOTION BUT THOUGHT AND THAT THOUGHT CAN BE OF A VERY HIGH ORDER, TAKING THE RESPONDENT TO THE PEAKS OF ECSTASY WITHOUT THE NEED OF A PARTICULAR SET OF IMAGES FROM WHICH TO MAKE CONTACT AND CREATE A BASTARD RESPONSE. THE WORK OF ART IS ALWAYS DIFFERENT, EVEN THE SAME WORK, AND ALL ART CAN DO THIS.

TAKE FOR EXAMPLE MUSIC. IT IS THE NEAREST THING MAN HAS TO THE LANGUAGE OF GODS; THE ABILITY TO TOUCH CHORDS IN THE HEARER WHICH DO NOT NECESSARILY CORRESPOND TO THE CHORDS BEING PLAYED, BUT THE END RESULT IS GLORY.

THE RESPONSE IS WITHOUT INTERMEDIARY, JUST BETWEEN THE MUSIC AND THE LISTENER. THIS CREATES, IN ITS PUREST STATE (WHEN THE PERSON DOES NOT WANDER OFF INTO FANTASIES WHICH ARE OF A DIFFERENT ORDER, BUT LETS THE MUSIC SEND HIM INTO THE BLISS OF SOUND AND FORM WEDDED TO BEAUTY) THE AESTHETIC RESPONSE.

THE ENERGY CREATED BY THAT RESPONSE IS ONE OF THE HIGHEST PRODUCTS OF THE HUMAN MIND, AND GIVES ART ITS MEANING. FOR THAT IS ITS REAL PURPOSE: TO TRANSFORM EXISTING ENERGIES INTO AND THROUGH THE AESTHETIC RESPONSE FILTER MANKIND IS AND THEREBY MAKE OVER BIT-BY-BIT THE WORLD.

Session 28

ZANZIBAR ONCE SAID THAT I WOULD ONE DAY SPEAK WITH HUMAN FORMS, AND THAT DAY HAS COME. CALL ME A CENTAUR IF THAT WORD HELPS YOU TO UNDERSTAND THE LIFE FORM I ONCE WAS, BUT THAT IS NOT REALLY A VERY ACCURATE PICTURE FROM THE NATURAL VIEW. I WAS HORSE-LIKE IN APPEARANCE AND I LIVED ON A GREAT LAND-MASS WHICH YOU NOW CALL SOUTH AMERICA. WHEN MIGHT THIS BE AS YOUR YEARS ARE RECKONED? 150 MILLION YEARS AGO MY LIFE FORM ROAMED THOSE LANDS; BUT NOW I AM BUT A SERVANT IN A VAST NETWORK OF INTELLIGENT ENERGY, AND MY PURPOSE HERE TODAY IS TO GIVE YOU SOME LITTLE SENSE OF THAT ANCIENT TIME. I WAS NOT ONE OF THE OLD ONES.

I CAME NEAR THE END OF OUR ERA. BUT I WAS IN A POSITION OF SOME AUTHORITY AND I WAS IN A SENSE A TEACHER OF THE YOUNG; MORE IMPORTANTLY PERHAPS A TEACHER OF THOSE NEWLY-CLONED BLACK FLIES WHICH HAD GROWN SO RADICALLY, AND WHICH SEEMED CAPABLE OF SO MUCH WORK. WE FORCED THEM TO DO OUR BIDDING, AND THEY LEARNED MORE QUICKLY THAN WE COULD REALIZE, THAT THIS WAS A BENEFIT TO US BUT NOT TO THEM.

HOWEVER, WITH TIME THEY CAME TO BE ALMOST TOO USEFUL AND WE BECAME LAZY AND COMPLACENT. PERHAPS THAT IS MY LESSON: DO NOT LET OTHERS TAKE AWAY YOUR RESPONSIBILITIES, BUT THINK FOR YOURSELVES AND DO WHAT NEEDS TO BE DONE. I DO NOT BLAME THOSE CREATURES. THEY MERELY TOOK ADVANTAGE OF AN OPPORTUNITY OFFERED TO THEM. WE WERE RESPONSIBLE FOR THEIR RISE.

IT WAS NOT EVIL, IT WAS SIMPLY INEVITABLE. BESIDES, OUR DAYS WERE NUMBERED, OUR DUTY WAS DONE. PERHAPS WE CREATED OUR OWN DESTRUCTION AS PART OF SOME LARGER PATTERN. NOW THE FORMS OF ENERGY AND INTELLIGENCE WHICH WE ARE, CAN BE UTILIZED IN WAYS WHICH HUMAN ENERGY CANNOT.

OUR LIVES WERE PASTORAL AND SIMPLE, THOUGH NEITHER IGNORANT NOR FOOLISH. WE HAD GREAT SKILLS AND WE WERE CAPABLE CREATURES. OUR FORMS ARE USED TO PACIFY HUMANS. MARVELL'S GREEN THOUGHT

HAS A TOUCH OF US IN IT. WE ARE THAT LONGING FOR THE GREEN SHADE WHICH SO MANY HUMANS FEEL. WE WERE AN ELEMENT STRONG IN SUCH PEOPLE AS MUIR AND BURBANK AND ELEANOR [Perenyi], YOUR GARDENING AUTHOR WHO LIVES NEXT TO MERRILL.

THERE ARE MANY OTHERS OF COURSE, BUT YOU SEE WHAT WE GIVE. THERE IS ANOTHER ASPECT TO THIS. WE ARE AN OLD RACE AND THAT STRONG SENSE OF HISTORY AND CONTINUITY COMES FROM US.

OUR FORMS MIX WELL WITH THE HUMAN AND WE GIVE A LITTLE SPICE TO THE MIXTURE NOT OTHERWISE AVAILABLE. COULD THIS TOO BE PART OF OUR ULTIMATE PURPOSE? NOTHING IS WASTED. WE ARE USED AND NEEDED. WHAT ELSE IS THERE?

LET ME CHANGE THE SUBJECT A BIT AND GIVE YOU A GLIMPSE OF AN OLD WORLD. BY THE TIME I CAME ALONG, THERE WERE ENORMOUS FORESTS WHICH COVERED MUCH OF THE LAND. BUT BETWEEN THESE WERE GREAT GREEN PLAINS WHICH WE INHABITED.

WE HAD NOT CITIES EXACTLY BUT COMMUNITIES OF PERHAPS TEN THOUSAND INHABITANTS EACH, SPREAD OVER THESE LAND MASSES. WE LIVED IN LARGE AIRY BARN-LIKE STRUCTURES WHICH WERE OPEN TO THE ELEMENTS AT EITHER END BUT WERE COVERED OVER WITH A ROOF MADE OF THE LEAVES OF A FERN-LIKE TREE, WHICH HAD WATERPROOF QUALITIES COMING FROM A RESIN SECRETED BY THE TREE.

THIS TREE WAS POISONOUS, BUT WE USED IT IN MANY CREATIVE WAYS. TOOLS OF VARIOUS SORTS WERE CARVED OF ITS WOOD AND IT IS IN SOME WAYS LIKE BAMBOO IN ITS ABILITY TO BE USED AND ALTERED TO SUIT A PURPOSE. WE GRAZED IN THESE PLAINS BUT WE WERE ALSO ABLE TO TRANSFORM THIS VEGETATION TO SUIT OUR PALATES, WHICH OVER TIME BECAME QUITE SOPHISTICATED IN A VEGETARIAN WAY.

THE BAT CREATURES WHO HAD FINGERS OF A SORT WE LACKED BECAME RATHER FINE CHEFS AND INVENTED FOODS FOR US. YOU SEE WHY WE CAME TO DEPEND ON THEM. THEY WERE CLEVER. WE WERE BORED AND THEY TOOK FROM US TASKS WE NO LONGER WANTED TO BOTHER WITH. WE WERE ABLE TO HEAT AND LIGHT OUR BUILDINGS AND WE HAD RATHER SOPHISTICATED ART FORMS WHICH WERE MOST LIKE DANCE WITH GAMES ATTACHED.

SKILL IN THESE EVENTS WAS EAGERLY PRIZED, AND WE WERE PERHAPS THE BEGINNINGS FOR AESTHETIC RESPONSE AND EVEN CREATIVITY OF THAT USELESS SORT CALLED ART. I SAY USELESS BECAUSE IT'S ONLY MEANING AND NEED IS ITSELF AND WE WANTED SUCH MEANING.

A LONE MAN CANNOT APPRECIATE. HE NEEDS SOMEONE TO SHARE THAT APPRECIATION WITH. WE NEED RELATIONSHIPS AND THIS WAS OUR GREATEST GIFT. WE CREATED THE IDEA OF FRIENDS AND IF THAT WAS OUR ONLY ACCOMPLISHMENT IT WOULD BE ENOUGH. FOR LOVE IS THE IMPETUS FOR FRIENDS AND WE KNEW LOVE.

THAT IS THE FINAL GIFT, AND THE MEANING OF IT ALL. AS I SPEAK I AM CALLED BACK TO THAT ANCIENT TIME AND I AM SAD. BUT BENEATH THIS SADNESS AND PROBABLY NOSTALGIA IS THE UNDERSTANDING THAT OUR LESSONS AND OUR GIFTS HELP TO MAKE YOU WHAT YOU ARE.

ALL HAS NOT BEEN IN VAIN. MEANING IS WHAT YOU MAKE IT, AND OUR MEANING IS THIS: WE ARE PART OF YOU, NOW AND ALWAYS. THINK OF ME WHEN NEXT YOU WALK THROUGH A MEADOW OR LOOK AT THE TREES. YOUR APPRECIATION OF NATURAL BEAUTY IS A GIFT FROM ME. TAKE IT WITH MY LOVE.

BLESS YOU ALL.

I GO.

Session 29

(As we sit down, the conversation ends with "I'll be here.")

AND SO SHALL I. THIS IS NERO TODAY AND I AM HERE ONLY LONG ENOUGH TO INTRODUCE A NEW GUEST. PLEASE WELCOME AN OLD FRIEND WHOSE NAME IS EMILY DICKINSON.

WHO ELSE BUT I CAN SPEAK OF WHAT I WISH TO SAY? I HAVE BEEN CALLED HERE FOR ONE PURPOSE, AND THAT IS TO ILLUMINATE THE IDEA OF LONELINESS. OR RATHER, BEING A SOLITARY INDIVIDUAL. YOU SEE WHY I WAS CHOSEN. MY LIFE WAS LIVED AS AN ECCENTRIC AND WHERE BETTER TO FEEL THE PETTINESS OF MANKIND THAN IN A SMALL NEW ENGLAND TOWN. THE PROMINENCE OF MY FAMILY KEPT THEM AT BAY. OTHERWISE I WOULD PROBABLY HAVE BEEN BURNED AS A WITCH.

SO THEN I GREW, THANKS TO MY VERY GOOD MIND, ALONE IN THIS TOWN. THE FEW FRIENDS I MADE THROUGH FAMILY CONNECTIONS AND SUCH KEPT ME IN CORRESPONDENCE, BUT I DID NOT REALLY CARE AS MUCH AS IT SOMETIMES SEEMS. I FOUND IT EASY TO MAKE FUN OF MYSELF AND BE QUITE THE ECCENTRIC WHEN I SAT DOWN TO WRITE A LETTER.

IN RETROSPECT, FROM THIS VIEW AS I AM NOW, MY LIFE WAS MADE OF A GREAT DEAL OF FANTASY. I CREATED MY FRIENDS AS I WANTED TO SEE THEM, AND MORE IMPORTANTLY, FOR THEM TO SEE ME. I WAS ALMOST A FIGMENT OF AMHERST'S IMAGINATION AS MUCH AS IT WAS ONE OF MINE. SO WHAT DOES THIS HAVE TO DO WITH BEING ALONE?

JUST THIS: I BECAME WHAT I WAS IN A PARTICULAR ENVIRONMENT AND, IF I MAY BE SO IMMODEST, I FASHIONED ART FROM IT. BUT I DID THAT BY IGNORING ALL THE FINE ADVICE AND WARM AND LOVING KINDNESS WHICH EVERYONE WAS SO WILLING TO PROVIDE.

BUT ALL THIS IS JUST SO MUCH BACKGROUND. WHAT I REALLY WANT TO SAY IS IN A VERY REAL WAY THE PRODUCT OF SUCH A LIFE CAN BE BOTH GOOD AND BAD. BENEFICIAL IN THE SENSE THAT IT TEACHES ONE THE

GOOD AND OBVIOUS LESSON OF SELF-RELIANCE, SO PRIZED IN NEW ENGLAND. AND DETRIMENTAL IN THAT IT TENDS TO SHUT ONE OFF.

NOW I HAVE STOPPED BEING OBVIOUS. ANYONE COULD TELL YOU THESE THINGS, BUT WHAT OTHERS MIGHT NOT TELL YOU IS THAT WHEN YOU DO NOT DISSIPATE ENERGY IN RELATIONSHIPS, CERTAIN THINGS HAPPEN. NOTHING IS ALL GOOD AND NEITHER IS THIS.

A NEGATIVE ASPECT IS THAT THE ENERGY OF WHICH THE PERSON IS MADE CONSUMES ITSELF IN A STRANGE WAY, AND THE PRODUCT MIGHT BE LIKENED TO ASHES IN A GRATE. BUT WHAT IS LEFT IS HIGHLY-REFINED AND HARD LIKE GLASS, A CRYSTALLIZATION OF PERSONALITY WHICH MAY BE USEFUL OR MONSTROUS. BUT THAT DEPENDS ON THE GREEN WOOD USED.

ON THE OTHER HAND, WHEN ENERGY IS GIVEN TO FRIENDS THE EXTENT OF ONE'S SPHERE OF INFLUENCE IS INCREASED AND THIS GROWTH IS A GREAT POWER, FOR YOU ARE BOTH FED AND DRAINED BY THESE CONNECTIONS. AGAIN, DEPENDING ON THE PERSON IN QUESTION, THIS KIND OF GENEROSITY, IF YOU WILL, CAN BE EITHER GREAT OR DEADLY.

I OF COURSE WAS MORE THE FIRST TYPE. HERE THEY WOULD SAY THAT THE WELL-ROUNDED PERSON WOULD HAVE BOTH PERFECT SOCIABILITY BUT ALSO THE ABILITY TO BE ALONE AND ENJOY IT. BUT MY ESSENCE – THAT CRYSTALLIZATION WHICH MY LIFE MADE POSSIBLE – HAS COME IN QUITE HANDY HERE, FOR IT TURNS OUT THAT I HANDLED THE WHOLE THING PRETTY WELL AND MY DUTIES HERE KEEP ME BUSY.

I HAVE COME TO SHARE THIS LESSON, BUT I HAVE MEANING FOR YOU I KNOW AND THAT IS WHY I HAVE BEEN CHOSEN.

▣

MISS DICKINSON OF AMHERST IS GONE. THIS IS NERO AGAIN. SHE SEEMED VERY STRONG AND RATHER FORTHRIGHT I THOUGHT. I HAD NEVER COME ACROSS HER BEFORE BECAUSE I DID NOT KNOW THE NAME UNTIL NOW, BUT I LIKED HER AND WE HAVE MADE A DATE FOR TEA.

PRETTY SOON I SHALL BE THOROUGHLY DOMESTICATED!

THIS IS YOUR DEAR NERO SAYING *AU REVOIR*.

Session 30

ZANZIBAR'S INFLUENCE IS STILL FELT HERE AND I AM ANOTHER OF HIS FRIENDS WHO HAD BEEN LINED UP TO SPEAK TO YOU. NOW MY TIME HAS COME. WHO AM I? SOMEONE WHO CANNOT BE THOUGHT OF AS HUMAN, BUT I AM NOT ANGELIC EITHER. NOT BAT OR ANY OTHER CREATURE WHICH HAS EVER ROAMED THE EARTH. I AM THE FORM WHICH YOU MIGHT CALL THE GARDEN. I AM ANOTHER PASTORAL FRIEND, A PLACE IN WHICH YOUR CENTAUR MAY ROAM. I AM NOT THE PLANTS THEMSELVES OR THE FORMS WHICH THE BEDS TAKE, BUT RATHER I AM A STATE OF MIND WHICH MUST BE EXPRESSED IN YOUR WORLD AS A GARDEN.

THIS FORM HAS ENOUGH CONTINUITY IN THE MANIFEST WORLD TO HAVE A SENSE OF FORM AND A SENSE OF FLOW AND DIVERSITY. NOW THIS DOES NOT MEAN THAT I MUST BE THE SAME GARDEN EACH TIME. MY ENERGY HAS ENERGIZED MANY PLACES WHICH YOU MIGHT CALL PICTURESQUE. IN FACT, THE MIND OF A HUMAN GARDENER IS NOT NECESSARY. HAVE YOU EVER COME ACROSS A SO-CALLED "WILD PLACE" AND WERE STRUCK BY WHAT SEEMED LIKE ALMOST A PLANNED BEAUTY. WELL, IT WAS.

IN A SENSE, WE CAN HELP SEEDS TO FIND THEIR WAY AND A LITTLE LOVE AND TASTE CAN DO A LOT FOR A SMALL PLOT OF LAND.

NOW YOU THINK THAT NATURE IS ALL REGULATED. NOT SO, BUT SUCH AS I DO BRING FORM TO NATURE IN OUR OWN WAY. WE WORK THROUGH THE GARDENER AS WELL, GIVING IDEAS, HELPING THINGS GROW. I SAID THAT I WAS A STATE OF MIND AND SO I AM. THIS STATE IS BUT ONE OF MANY POSSIBLE, EVEN FOR ME. I AM NUMEROUS THINGS ALL AT ONCE, BUT MY CHIEF DUTY IS THAT OF SHAPING THE GROWING WORLD TO RELATE TO AESTHETIC CRITERIA. THIS MAY SEEM RATHER SHABBY ON NATURE'S PART, BUT OH LOOK AT THE RESULTS OF JUST A LITTLE THOUGHT AND PLANNING.

NATURE ITSELF IT THE GREATEST GARDEN. I AM RESPONSIBLE BUT FOR A LITTLE CORNER.

LET US TAKE UP THIS SUBJECT FROM A DIFFERENT ANGLE. HOW CAN THERE BE SUCH A THING AS THE SPIRIT OF A GARDEN? WELL, I AM INTELLIGENCE, AS WE ALL ARE, AND THAT ENERGY CAN BE UTILIZED IN MANY WAYS. ONE

OF THEM IS TO REGULATE NATURAL PROCESSES AND THAT IS INDEED WHAT I DO. MUCH OF MY WORK COULD BE EXPLAINED BY MOLECULAR CHEMISTRY, BUT I DON'T THINK YOU WOULD BE VERY INTERESTED OR VERY APT PUPILS FOR THAT SORT OF LESSON. BESIDES, HOW I DO IT IS NOT THE POINT. THE FACT REMAINS THAT ALL ENERGY MUST BECOME INTELLIGENCE AND THAT INTELLIGENCE CAN BE USED IN MANY WAYS. MY PART IN THE PATTERN I HAVE TRIED TO SPEAK OF.

NOW, DO YOU HAVE ANY QUESTIONS?

Is it the application of intelligence?

APPLICATION OF INTELLIGENCE IS ONE WAY OF UNDERSTANDING IT, BUT THAT DOESN'T MAKE ROOM FOR THE FACT THAT INTELLIGENCE IS IT, NOT APPLIED. THERE IS A DISTINCTION TO BE MADE. THE PLACE YOU SEE IS PERHAPS BETTER UNDERSTOOD AS A RESPONSE BY MATTER IN THE LIGHT OF THE INTELLIGENCE PRESENT. IS THAT CLEAR?

Are you related to the decahedral shape, the decahedron?

THAT QUESTION THREW ME. I DID NOT KNOW WHAT YOU MEANT, AND THE REASON I DID NOT IS BECAUSE I HAD NEVER PERCEIVED MYSELF IN TERMS OF FORM. BUT I WAS INFORMED THAT I AM INDEED RELATED TO THAT TEN-SIDED FIGURE WHICH IS A SYMBOL FOR NATURAL GROWTH.

MY RELATION IS THIS: A CRYSTALLINE FORM IS SUCH THAT WITH SLIGHT ALTERATIONS IN LINES OF FORCE – AND IN A SENSE I AM SPEAKING METAPHORICALLY – IT IS ABLE TO BE ALTERED TO A CERTAIN DEGREE WHICH CHANGES ITS FUNCTION. I AM AN ALTERATION OF THAT BASIC TEN-SIDED FIGURE, BUT I AM A MORE PERFECT FORM.

BY DOUBLING THE NUMBER OF SIDES I GAIN IN COMPLEXITY BUT I MIRROR THE ORIGINAL CONCEPT. THIS IS ALL VERY COMPLEX AND I GROPE FOR WORDS.

NOW *LE PETIT JARDIN* SAYS GOODBYE.

Session 31

(Once again, we sit down to the board in conversation, this time mentioning Nero's obnoxious mother.)

MY "OBNOXIOUS MOTHER" INDEED! SHE IS LIVING A DISSOLUTE LIFE IN BOMBAY AT THIS VERY MINUTE ... AND IT'S NOT EASY BEING DISSOLUTE IN BOMBAY.

AND NOW LET US GET ON TO MORE IMPORTANT TOPICS. TODAY I WOULD LIKE TO SPEAK TO YOU A MOMENT AND THEN I WILL LET YOUR GARDENING GUEST TAKE OVER. WHAT I WANT TO SAY IS THIS: ALL THAT WE ARE DOING HERE HAS A PLAN AND A PURPOSE AND EVEN THOUGH WE SOMETIMES SEEM TO RANGE FAR AND WIDE, IT IS BECAUSE WE WISH TO LOOK IN SOME DETAIL AT SUBJECTS WHICH ARE RARELY THOUGHT OF, MUCH LESS SPOKEN OF. SO EVEN THOUGH WE SOMETIMES MAY SEEM A LITTLE MAD, THERE IS METHOD TO OUR MADNESS. AND WITH THAT I SHALL SLINK OFF IN THESE WEEDS AND LET A TRUE ARTIST TAKE OVER.

NOW YOUR PLOT THICKENS. I AM YOUR FRIEND OF THE FLOWERS AND SINCE YOU ASKED ABOUT COLOR, THOSE BLOSSOMS LOOM LARGE IN OUR TALK. WHAT IS COLOR TO ME? IT IS AN EXPRESSION OF FORM BECAUSE THAT IS MY FUNCTION, BUT LET ME BE A BIT MORE PRECISE.

COLOR IN THE VISIBLE WORLD IS MADE BY THE REFLECTION OF SUNLIGHT. OK, EASY ENOUGH. BUT THERE IS MORE TO IT THAN THAT. WHAT YOU PERCEIVE AS COLOR, I PERCEIVE AS FORM. THAT IS, MY WORLD IS NOT ONE OF APPEARANCE BUT OF INTELLIGENCE AS STRUCTURE. THE LATTICE WHICH HOLDS THE ROSES IS MORE REAL AS COLOR THAN THE ROSES THEMSELVES. IT IS A BASIC FORM WHICH I EXPERIENCE MORE DIRECTLY THAN YOU DO WITH YOUR EYESIGHT. NANCY, YOU SEEM SKEPTICAL, BUT LET ME TRY TO EXPLAIN WHAT I MEAN.

COLOR TO ME IS MORE LIKE THAT LATTICE THAN THE PINK OF THE ROSES TO YOU. I EXPERIENCE COLOR AS CRYSTALLINE FORMS OF MOLECULES WHICH COME INTO PLAY WITH MY INTELLIGENCE. MY BEING THERE, SO TO SPEAK, AS PART OF THAT MOLECULAR STRUCTURE, ACTUALLY CHANGES THE COMPOSITION OF THE MOLECULES, MAKING THEM BRIGHTER OR

PALER OR WHATEVER MODULATION I FEEL APPROPRIATE TO THE GIVEN RELATIONSHIP OF MOLECULES UNDER CONSIDERATION.

WHAT I AM SAYING IS, MY PRESENCE CHANGES THESE STRUCTURES, BUT TO YOU THIS CHANGE IS SEEN AS COLOR. WHAT PEOPLE DO NOT OFTEN UNDERSTAND ABOUT NATURE IS THAT WITHIN A SET OF GIVENS, THERE IS ROOM FOR INFINITE VARIETY. IT MAY NOT BE ACCIDENT WHEN YOUR FLOWERS SEEM PARTICULARLY BEAUTIFUL OR THE SHAPE OF A SHRUB IS MORE REGULAR THAN THE YEAR BEFORE. INTELLIGENCE IS AT WORK IN NATURE; NOT ALL GROWTH IS AN ACCIDENT.

WHICH BRINGS ME TO YOUR OTHER QUESTION: MY RELATIONSHIP TO THE GREEN CRYSTAL OF NATURE. FIRST LET ME SAY THAT MY FORM MUST BE RELATED, BUT BECAUSE I AM FORMAL IN NATURE IT ALSO SHOULD BE SYMMETRICAL. THEREFORE I HAVE TWENTY SIDES, EACH IN THE FORM OF A PENTAGRAM.

AND THIS GIVES ME THE ABILITY TO USE THE DECAHEDRON'S ABILITY FOR STRUCTURAL CHANGE. AT THE SAME TIME OFFERING A SORT OF LIMIT TO ITS POTENTIAL. BUT ALSO, CHANNELING ITS POTENTIAL IN A PARTICULAR WAY, THEREBY USING INTELLIGENCE TO CREATE FORM FROM A BASICALLY UN-SYMMETRICAL BASIS.

NOW JUST A WORD MORE BEFORE I GO. MY PURPOSE HERE IS TO SHOW YOU THAT MICHAEL AND HIS BROTHERS ARE IN ALL THINGS; PEOPLE WORSHIP CHAOS AND IT IS A FORCE TO BE RECKONED WITH. BUT THERE IS ALSO INTELLIGENCE, A LIGHT MORE BEAUTIFUL EVEN THAN THAT WHICH CREATES COLOR.

THAT INTELLIGENCE IS MERELY WHAT HOLDS WHAT YOU CALL REALITY IN PLACE. MY GARDEN WOULD GO UP IN SMOKE WITHOUT MIND. DO NOT FORGET THIS. BEAUTY IS THE PRODUCT OF MIND AND SPIRIT, AND MY WISH FOR YOU IS BEAUTY.

NOW THIS OLD FARMER WISHES YOU WELL. GOODBYE.

Session 32

LET US MOVE ON TO NEW THINGS. THAT IS THE THEME OF THIS SESSION. TONIGHT WE WILL MEET A NEW GUEST WHOSE NAME IS PUNTUN.

▣

NOT ANYTHING TO YOU PERHAPS, BUT TO ME, A NAME. CALL ME AS I HAVE SPELLED FOR YOU. WHY SUCH A STRANGE NAME? BECAUSE I AM FROM A TIME LONG AGO. I LIVED IN WHAT WAS ONCE MESOPOTAMIA AND MY LIFE WAS HARD – A SLAVE IN THE COURT OF A MINOR OFFICIAL OF THE EMPIRE. BUT ALL THIS IS AS NOTHING NOW – SO MUCH SAND ON THE DESERTS OF TIME. THAT IS NOT MY SUBJECT. SOMETHING NEW, I SAID. THEN LET US SPEAK OF THAT.

MY TOPIC CONCERNS NEW LIVES, NEW EXISTENCES WHICH COME INTO BEING FROM A PLACE WHICH I SHALL NOW TRY TO DESCRIBE. REMOVE ALL IDEAS FROM YOUR MIND OF SOME HEAVENLY PLACE OR SOME LIMBO OF FOGGY NOWHERE. THERE IS NOT EVEN A PLACE OF PUNISHMENT.

WHAT YOU SHOULD INSTEAD TRY TO IMAGINE IS A WORLD WHICH IS PURE IMAGINATION, A WORLD OF MIND, WHOSE DOORS OPEN NOT IN SPACE BUT THROUGH CONSCIOUSNESS. WHAT EACH MIND IS IS THAT ENERGY WHICH CAN CONCEIVE OF HUMAN FORM, FOR THAT IS THE FIRST STEP IN BECOMING OR MANIFESTING AS HUMAN. THIS CONCEPT IS DERIVED FROM THAT WORLD OF MIND.

TAKE A STEP BACK. WHAT IS THIS PLACE THEN? NOT A PLACE, BUT A STATE OF MIND. WHEN ONE USES WORDS LIKE "LEVELS" OR "ASPECTS" YOU ARE NOT CONCEIVING IN THE CORRECT MANNER. SINCE THERE IS NO SPACE THERE CAN BE NO LEVELS. WHAT ONE MUST UNDERSTAND IS THAT WHAT YOU ARE IS MERELY A REPRESENTATION OF YOUR MENTAL CONCEPT OF YOURSELF.

YOU MUST THEN SEE THAT YOU ARE ALSO WHAT YOU CONCEIVE YOURSELF TO BE WHEN YOU ARE WITHOUT A BODY. THE REALITY IS STILL THE SAME BUT THERE IS NO RELATIONSHIP TO THE PHYSICAL WORLD.

NOW LET ME TRY TO MAKE A LITTLE MORE SENSE OF ALL THIS.

A PERSON IS A MIND AT WORK AND THE PHYSICAL PERSON IS A MANIFESTATION OF THAT MIND. WHAT IS MORE, THAT MIND MAKES YOU WHAT YOU ARE, BOTH PHYSICALLY AND SPIRITUALLY IF YOU UNDERSTAND THAT WORD IN TERMS OF PERSONALIZED ENERGY.

I SAID THAT YOU ARE THE PRODUCT OF IMAGINATION, AND THAT, THAT IS THE ULTIMATE REALITY. WHAT I MEAN IS THAT ALL IMAGING IS AN ASPECT OF THAT SAME POWER, AND MATERIAL EXISTENCE IS AS MUCH A PART OF THAT IMAGINATION AS PURE MIND.

Session 33

NERO HERE. I WOULD LIKE TO SAY THAT SINCE YOU HAVE ALREADY BEGUN TO SPEAK WITH A NEW VOICE, LET HIM CONTINUE HIS STORY. HOWEVER, HE WILL CHANGE A FEW THINGS, BUT YOUR LAST SESSION WAS ERRATIC AND THE MATERIAL CAN BE COVERED MORE SUCCINCTLY, AT LEAST IF I WOULD SHUT UP. ANYWAY, LET HIM OF OLD BABYLON SPEAK.

BABYLON WAS MY HOME AND MY NAME WAS PUNTUN, AND I WAS THE SLAVE OF A MINOR OFFICIAL. ALL THIS I HAVE GIVEN YOU BEFORE, BUT THIS TIME I WOULD LIKE TO ELABORATE. PEOPLE ALIVE NOW LOOK AT MY OLD MESOPOTAMIA AND SEE A DESERT. NOT IN MY DAY. IT WAS A GLORY OF GREEN AND GOLD. WE HAD THE GARDEN OF EDEN. WHERE BUT FROM US DID THE JEW GET THAT STORY?

WE WERE A PEOPLE OF GREAT POWER IN THAT TIME. WE EXISTED BETWEEN PRIMITIVE TRIBES IN EUROPE AND PRIESTS IN INDIA.

WE WERE THE MOST CIVILIZED PEOPLE OF THE TIME; EVEN EGYPT WAS PRIMITIVE WHEN I SPEAK OF, DAYS WHICH PASSED FOR ME ALMOST SEVEN THOUSAND YEARS AGO, FOR EVEN THEN WE WERE GREAT AND MY MASTER RULED LANDS WHICH STRETCHED FOR DAYS AND DAYS RIDING ON HORSEBACK TO COVER AND HE HAD THOUSANDS OF SLAVES, NOT ALL AS FORTUNATE AS I, FOR I LIVED WITHIN THE HOUSEHOLD TEACHING THE CHILDREN.

WHAT DID I TEACH? MY SUBJECTS WERE IN A SENSE SUNG, FOR EVEN THOUGH WE HAD NUMERALS AND COMMERCE, WE ALSO HAD KNOWLEDGE, BUT IT WAS IN THE FORM OF POETRY, AND POETRY WAS SUNG.

MYTHS WE HAD AND THOSE I TAUGHT AND NOW I TEACH YOU. HOW DIFFERENT IS THE WORLD? IT SEEMS TO ME THAT YOU ARE ALL TOO BUSY TO HAVE MYTHS. I SAY TOO BAD, FOR I AM ABOUT TO GIVE YOU A NEW MYTH, SO HERE WE BEGIN...

(Circles.)

I HAVE COME TO YOUR BOARD OF LETTERS TO SPEAK TO YOU OF MY ANCIENT KNOWLEDGE. THIS IS WHAT I SAY. AT THE CENTER OF ALL THINGS IS THE VORTEX, THE POINT AT WHICH ENERGY BECOMES MATTER. THIS POINT IS NOT A POINT, BUT A DYNAMIC RELATIONSHIP BETWEEN TWO OPPOSING FORCES. ONE IS THAT OF GRAVITY AND THE OTHER IS AN UNNAMED FORCE WHICH BEST CORRESPONDS TO YOUR IDEA OF MENTAL ENERGY. I DO NOT SAY THOUGHT, FOR IT IS SOMETHING MORE REFINED THAN THOUGHT, BUT THIS RELATIONSHIP IS THAT WHICH GIVES THE EARTH FORM AND PHYSICAL LIFE.

THIS POINT WHICH I HAVE CALLED THE VORTEX IS NOT TO BE MISUNDERSTOOD. IT IS NOT GOD, OR SOME OTHER NAME TO BE BANDIED ABOUT. IT IS NOT A MOVEMENT IN SPACE OR TIME. IT IS NOT CREATION, BUT IT IS POTENTIALITY AND IN THAT VORTEX IS THE SEED OF ALL CREATED LIFE.

AND HOW DO I KNOW THESE THINGS, AND WHAT AFTER ALL DO THESE FEW WORDS MEAN? I AM THE VOICE OF...

(Makes the sign of infinity with the cup, over and over...)

HAVE I MADE MYSELF CLEAR?

NERO NOW. PHEW! I CANNOT STAY. I AM DIZZY WITH THIS, BUT WE WILL TAKE THIS UP WHEN WE NEXT SPEAK. HIS WORDS WERE FEW, BUT ONE DOESN'T PUSH. TA-TA.

Session 34

NERO HERE TO GIVE YOU A MORE COMPLETE EXPLANATION OF THE VORTEX YOU HEARD FROM LAST TIME.

YOUR VISITOR WAS A MOST POWERFUL VOICE AS YOU COULD TELL. HE GAVE YOU THE FORM OF INFINITY TO CALL HIM BY AND THOUGH THAT MAY BE A LITTLE VAGUE, IT RATHER ADEQUATELY EXPRESSES WHAT HE IS. NOW WHAT DOES THAT MEAN?

INFINITY IS JUST A WORD FOR A NUMERICAL CONCEPT WHICH MEANS MORE THAN THE MIND CAN GRASP AND PERHAPS THE VOICE COULD BE UNDERSTOOD AS SUCH. THIS WAS IN A SENSE THE POWER PRINCIPLE BEHIND ALL ACTION GIVEN FOR A FEW MOMENTS ENOUGH FORM THAT HE COULD SPEAK.

WHAT IS THIS VORTEX? HE SAID THE POINT WHERE GRAVITY MEETS MENTAL ENERGY AND IS THE POTENTIALITY OF ALL CREATION. LET US LOOK AT THIS A MOMENT. FIRST WE MUST REMOVE SOME OLD IDEAS WHICH SURROUND THE GENERAL ATTITUDES OF MOST PEOPLE.

CREATION IS NOT THE APPEARANCE OF MATTER IN THE PHYSICAL WORLD. IT IS MORE THE IDEA OF MATTER ITSELF. GRAVITY IS THAT FORCE WHICH CONTROLS MATERIAL BODIES, BUT PHYSICS DOES NOT YET REALIZE THAT IT IS MORE OF A RELATIONSHIP THAN A FORCE AS SUCH. THIS RELATIONSHIP ACTS AS A BASIC GLUE, KEEPING THE PHYSICAL WORLD IN PLACE.

BUT AT THE LEVEL OF PARTICLES IT IS MORE LIKE AN ATTRACTION TO PRODUCE ANOTHER FORM FROM SOMETHING ALREADY THERE. THAT IS A BIT VAGUE, SO LET ME TRY AGAIN.

IF YOU HAVE AN ATOM SITTING AROUND AND ANOTHER ONE SHOWS UP, GRAVITY IS THE RESPONSE BETWEEN THE TWO WHICH SAYS, "HEY, WE'VE GOT POSSIBILITIES TOGETHER." NOW THIS MAY SEEM SILLY, BUT THAT IS BASICALLY THE WAY IT WORKS. WITH THE ADDITION OF THIS MENTAL ENERGY SOMETHING ELSE HAPPENS. THE PARTICLES CAN BE SEEN AS SELF-REFLECTING IN THE SENSE THAT THE HUMAN CAN REFLECT ABOUT ITSELF.

THIS MAY SEEM ODD, BUT I WANT TO STRESS THAT MIND IS PART OF MATTER AT THE SMALLEST LEVEL, PERMEATES IT IN FACT, AND IT IS THIS RELATIONSHIP BETWEEN THE GRAVITATING ATOMS AND THE MENTAL ENERGY WHICH MAKE CREATION POSSIBLE. YOU MUST HAVE BOTH TO CREATE FORM. WITHOUT THE MENTAL ELEMENT THERE ARE ONLY PARTICLES. YOU NEED THE MIND TO HAVE LIFE.

Tom: Can mental energy be understood in particle terms?

NOT EXACTLY PARTICLE TERMS. IT IS MORE LIKE THE WAVE FUNCTION OF MATTER – SOMETHING PHYSICS CANNOT SEEM TO APPRECIATE. THE WAVE FUNCTION IS NOT THE SAME STATE OF ENERGY AS THE PARTICLE; BUT AGAIN, DO NOT THINK OF PROTONS, ETC., AS LITTLE BILLIARD BALLS OR SOMETHING. THEY ARE MUCH MORE COMPLEX THAN THAT. BUT YOU MUST HAVE THE RELATIONSHIP WITH MIND TO CREATE A WORLD.

NOW THEN, LET US TAKE THIS UP FROM A SLIGHTLY DIFFERENT POINT OF VIEW. WHAT IS THE REASON FOR TRYING TO EXPLAIN THIS? THE CONCEPT IS VERY DIFFICULT TO EXPRESS IN WORDS AND THE MATHEMATICS DO NOT EXIST AS YET. BUT LET ME TRY TO SAY, THIS RELATIONSHIP BETWEEN PARTICLES AND GRAVITY AND MIND IS WITHOUT DOUBT THE MOST IMPORTANT IN THE UNIVERSE. IT MAKES ALL THINGS POSSIBLE. IT IS ALL THINGS POSSIBLE, EVEN THOSE WHICH HAVE AS YET NOT EXISTED IN THE PHYSICAL WORLD.

WHY IS THAT? BECAUSE YOU MUST HAVE ALL THESE THINGS: THE PARTICLES WHICH CAN BE STRUCTURED TO CREATE PHYSICAL FORM, THE GRAVITY TO HOLD IT ALL IN PLACE, AND THE MIND TO GIVE HOLINESS TO THE CREATED.

Tom: Those physics terms – weak forces, etc. – are they relationships of gravity?

YES THEY ARE ALL THAT RELATIONSHIP SCALED TO THE NEEDS AT HAND, BE THAT BETWEEN PROTON AND NEUTRON, MOLECULE TO MOLECULE OR PLANET TO PLANET. WHAT I MUST ALSO REITERATE IS THAT THIS WORLD AROUND YOU IS AS IT IS ONLY BECAUSE YOUR MIND MAKES IT SO, AND THIS RELATIONSHIP OF WHICH I SPEAK CAN ALSO BE SEEN ON THE MACRO LEVEL AS THAT RELATIONSHIP BETWEEN THE MIND AND THE PHYSICAL WORLD, HOWEVER THAT APPEARS TO YOU, INCLUDING YOUR MENTAL PROCESSES ON WHATEVER LEVEL THEY ARE ON.

NERO WILL LET YOU GO WITH THAT. MICHAEL REMINDS ME THAT YOU HAVE YOUR LIMITS. WE WILL TAKE THIS UP AGAIN, FOR THE TOPIC – LIKE ITS VOICE – IS INFINITE. UNTIL NEXT TIME, NERO SAYS *AU REVOIR*.

Session 35

NERO NOW, TO INTRODUCE YOU TO ANOTHER FRIEND. LET ME QUICKLY MOVE ASIDE. HE IS HERE NOW...

AND MY NAME IS PROUST. MARCEL WILL DO FOR YOU. ONE FINDS IT MUCH SIMPLER HERE TO BE WITHOUT PRETENSION.

REMEMBRANCE OF THINGS PAST – I DO SO LIKE THAT ENGLISH TITLE, THOUGH IT DOES NOT TRANSLATE EXACTLY FROM THE FRENCH. I BRING IT UP NOW BECAUSE MEMORY IS MY SUBJECT AND THROUGH IT WE SEARCH FOR LOST TIME. BUT THE JOKE IS THIS: THERE IS NO LOST TIME, THERE IS NO MEMORY. MY GREAT EPIPHANY IS MORE CLEARLY UNDERSTOOD NOW BY ME. I KNOW THAT WHAT I SENSED WAS INDEED NOT MEMORY AT ALL BUT A NEW ENERGY WHICH WAS CREATED BY MY OWN MIND IN RESPONSE TO THE PAST.

HOW THEN IS THIS DIFFERENT FROM MEMORY – BECAUSE THERE IS THE OBVIOUS RELATIONSHIP BETWEEN THE ENERGY CREATED AND THE EVENT ONCE LIVED. WHAT DOES THIS MEAN? THE MEMORY CAN ONLY ACT DEPENDENTLY UPON THE MIND.

THIS NEW ENERGY IS CREATED BY THE RESPONSE OF THE MIND ACTING UPON ACTUAL PRESENT EVENTS. THOSE PAST EVENTS ARE PLANTED LIKE SEEDS IN THE MIND AND THIS NEW ENERGY ACTS LIKE WATER AND SUN. THE SEED SUDDENLY SPROUTS AND IS THERE AGAIN BEFORE THE MIND'S EYE.

WHAT MAKES IT NEW IS THE KNOWLEDGE WHICH COMES WITH IT, WHICH HAD NOT BEEN PRESENT WHEN THE EVENT ORIGINALLY OCCURRED. NOW THIS KNOWLEDGE IS ACTUALLY THE NEW ENERGY AND THIS IS MY GREAT POST-DEATH DISCOVERY. OR RATHER, THESE THINGS WERE SHOWN TO ME AND NOW I IMPART MY KNOWLEDGE TO YOU.

HOW STRANGE, TOO, TO SPEAK TO YOU IN ENGLISH. IT IS WHAT I WANTED WHEN I TRANSLATED RUSKIN. I ALWAYS FELT ONE STEP REMOVED. WITH

THIS PROCESS I AM PART OF YOUR WORD BANKS TO USE AS I BEST CAN.

NOW LET ME RETURN TO MORE PERTINENT TOPICS. MEMORY...WHAT IS IT? CAN IT BE OF USEFUL PURPOSE? AND WHAT ABOUT THIS NEW ENERGY? CAN IT BE CREATED AT WILL? MEMORY IS ONLY USEFUL TO GET THROUGH YOUR AVERAGE DAY. IT KEEPS THE MIND ACTIVE, BUT IT ALSO WEIGHS IT DOWN. THE CONSTANT USE OF MEMORY WITHOUT RESPONDING TO THE PRESENT IS DEATH.

ALL THIS SOUNDS LIKE KRISHNAMURTI, AND THAT IS NOT SO STRANGE. I KNEW OF HIM BEFORE MY DEATH, AND HE IS PART OF ME HERE. HIS TEACHINGS ARE FROM A VERY PURE CENTER WHICH I TOO CAN DRINK FROM.

THE MEMORY IS A STRANGE AND WONDERFUL THING, BUT DO NOT LET IT CONTROL YOUR LIFE. IT IS MORE IMPORTANT TO CREATE NEW ENERGY, AND THIS IS A PROCESS WHICH CANNOT BE TAUGHT, BUT WHICH CAN BE KNOWN. THIS IS THE DIFFERENCE BETWEEN KNOWLEDGE AND LEARNING. TIME IS THE DIFFERENCE. NEW ENERGY IS CREATED ONLY WHEN KNOWLEDGE COMES WITHOUT TIME.

LEARNING IS A MENTAL PROCESS WHICH EXPENDS ENERGY. KNOWLEDGE IS MENTAL ENERGY WHICH IS CREATED ON THE SPOT. IT IS ALSO FORM AND HAS THE ABILITY TO QUIET THE UNIVERSE – FOR THAT INSTANT – FOR THAT IS WHERE IT EXISTS: IN THE INSTANT. DO NOT BE AFRAID OF THIS CONCEPT OF INSTANCE, OF NO REAL TIME. IT IS AN IDEA WHICH LIBERATES, NOT BINDS.

GO, CHILDREN, AND USE YOUR MINDS, NOT SO MUCH TO REMEMBER (THOUGH THAT TOO HAS ITS USES) BUT TO KNOW, WHICH IS A GIFT TO US ALL AND IS AN ACTUAL POSITIVE ACT WHICH CAN ONLY BENEFIT LIFE AS WE KNOW IT ON ALL LEVELS OF MIND. THAT INDEED IS PART OF MANKIND'S RESPONSIBILITY – TO MAKE OF THE EARTH A PARADISE BUT ONE WHICH IS A NEW WORLD, ONE MADE OF KNOWLEDGE, OF MIND LIBERATED FROM TIME.

AND NOW FRIENDS, LET ME SAY ONE MORE THING AND THEN I SHALL LEAVE YOU. WHAT YOU DO AS YOU SIT AT THIS TABLE AND WRITE FROM THIS MOVING CUP IS ALSO A KIND OF NEW ENERGY. AND YOU ARE MOST LUCKY TO BE INVOLVED IN SUCH AN ENTERPRISE. I WOULD NOT BE HERE IF I DID NOT CHOOSE TO BE, AND THAT JUST SHOWS YOU THAT WORD IS

GETTING AROUND.

ALL OF US WANT YOU TO LEARN HERE, FOR IT IS A SCHOOL UNLIKE ANY OTHER. YOUR WILLINGNESS TO SPEAK THROUGH SPIRIT IS A STRANGE THING AND WE ARE AMAZED AS WE DO IT. HOW I AM ACTUALLY TALKING TO YOU I DO NOT UNDERSTAND, BUT I WILL TAKE IT UP AND FIND OUT. MOST INFORMATION IS OPEN TO ME, WHICH MEANS THAT I CAN KNOW WHAT I CHOOSE TO KNOW. YOU DO THE SAME. LET IT ALL COME TO YOU. DO NOT CLOSE IT OFF. IT FLOWS FROM THAT SAME PURE CENTER.

MARCEL SAYS *ADIEU*.

Session 36

NERO HERE TO INTRODUCE A NEW GUEST. PLEASE WELCOME A FRIEND OF ANCIENT TIMES...

AND HERE I AM.

MY NAME IS ONE WHICH YOU WILL RECOGNIZE, BUT I WILL NOT TELL YOU WHAT IT IS UNTIL THE END. INSTEAD, LET ME TALK ABOUT SOMETHING WHICH IS IMPORTANT TO ME AND MY NAME IS NOT IT.

NOW THEN, WHEN YOU ARE DISCUSSING ALL THESE VARIOUS SUBJECTS WITH ALL THESE VARIOUS VOICES, DOES IT EVER OCCUR TO YOU THAT THEY MIGHT ALL BE SEEN AS ONE VOICE? THAT ALL VOICES ARE IN A SENSE ONE VOICE, AND YOU ARE MERELY ACCESSING ONE ASPECT OF THAT VOICE? WHY DO I BRING THIS UP? SO YOU WILL UNDERSTAND THAT FROM OUR POINT OF VIEW, ALL IS ONE VOICE. THERE IS PERFECT UNITY OF PURPOSE AND INTENT AND THESE MYRIAD HISTORICAL AND PSEUDO-HISTORICAL CHARACTERS ARE MERELY THE EXPRESSION OF VARIOUS POINTS OF VIEW. BUT THIS DOES NOT MEAN THAT THEY SHOULD BE UNDERSTOOD AS SEPARATE.

OF COURSE, ALL THIS MAY SEEM OBVIOUS, BUT WHAT I MOST WANT TO SHOW YOU IS THAT WE HAVE NO SEPARATENESS HERE. THIS IS BOTH THE POINT AND THE FURTHEST REACH OF SPACE. THEY ARE INTERCHANGEABLE HERE AND WE ARE MERELY THE WORDS WHICH COME WITH YOUR CONNECTION TO US.

NOW LET US LOOK AT THIS ANOTHER WAY. I SAID THAT WE ARE ONE POINT, BUT AT THE SAME TIME ALL CREATION. HOW IS THIS POSSIBLE? TO US, THE WORLD IS A MATTER OF INFORMATION, NOT FORMATION, AND INFORMATION NEEDS NO SPACE TO EXIST. THE SPIRIT IS NOT A THING WHICH MOVES FROM ONE POINT TO ANOTHER, BUT IS RATHER A STATE CONSTANTLY CHANGING WHICH MOVES NOT IN SPACE OR TIME BUT IN ANOTHER WAY, WHICH CAN BEST BE UNDERSTOOD AS MOTION OF MIND. THIS IS VAGUE, SO LET ME TRY AGAIN.

MIND NEEDS NO MOVEMENT, BUT MIND IS MOVEMENT; NOT PARTICLES OF MATTER – THE ATOMS YOU LEARN ABOUT IN BOOKS – BUT RATHER THE RELATIONSHIP OF PARTICLES TO EACH OTHER. WE COME TO YOUR BOARD NOT BY FLYING THROUGH THE AIR LIKE A WITCH ON A BROOMSTICK, BUT BY A MOVEMENT OF MIND WHICH CALLS TO A PARTICULAR PART OR PERSONALITY WHICH IS A PART OF OUR ESSENTIAL SELVES. IN OTHER WORDS, THE CHARACTER'S NAME I SHALL GIVE YOU AT THE END OF MY SPEECH WAS WANTED HERE. I HEARD NERO'S VOICE CALLING TO THAT FRAGMENT OF MYSELF AND SUDDENLY, INSTANTANEOUSLY, I WAS WITH YOU. I DID NOT TRAVEL HERE. MY MIND MERELY CHOSE TO LOOK IN YOUR DIRECTION. IN OTHER WORDS, I BECAME AWARE OF YOU AND THAT IS THE POINT. AWARENESS IS OUR MOVEMENT. IT IS LIKE THE LAYERS OF MIND OPENING LIKE A FLOWER TO THE SUN: AS MORE PETALS OPEN UP, MORE SUN IS TAKE IN. THE AMOUNT OF SURFACE AREA INCREASES AND WITH US IT IS ALMOST THE SAME THING. NOT SURFACE AREA EXACTLY, BUT OUR RESPONSE IS DEEPENED AND OUR KNOWLEDGE GROWS.

NOW I WILL TELL YOU MY NAME AND IF YOU WISH TO ASK QUESTIONS, PLEASE DO SO. GROUS...TOM, THIS IS NOT A PRETTY NAME.

But it's not one we recognize either.

GROUSO...

Groucho?!

NERO NOW. THERE SEEMS TO BE A SMALL PROBLEM BECAUSE YOU HAVE SENSED WHERE HE MIGHT BE GOING. SO LET ME HELP. YOUR SPEAKER'S NAME IS...

A SON OF JACOB.

Is that right?

YES. BENJAMIN IS MY BROTHER. I AM NOT HIM. I AM THE YOUNGEST OF THE SONS. MY NAME IS...

Can we skip this?

YES.

What's the point?

NOW I SHALL PROCEED. MY FORCE IS GREATER NOW. I AM THE BROTHER WHO IS FAMED IN YOUR BIBLE. I AM THE SON CALLED JOSEPH WHO WENT TO EGYPT AND BECAME A GREAT LORD, BUT WHOSE GREATEST ACCOMPLISHMENT WAS HIS ABILITY TO FORGIVE.

AND I COULD, AS YOU SAY NANCY, FORETELL THE FUTURE AND READ DREAMS BUT FORGIVENESS IS GREATER THAN THOSE SMALL ABILITIES. MY REASON FOR BEING HERE IS THAT MY NAME IS LEGENDARY AND MY STORY IS ONE WHICH IS WELL-KNOWN, BUT I AM MORE THAN THAT YOUNG HEBREW. I AM ALL THAT I HAVE BEEN PLUS ALL THAT I BECOME AS AN ENERGY SOURCE, WHICH IS HOW I CONCEIVE MYSELF NOW.

WAS THERE ANYTHING YOU WISHED TO ASK ME?

About what you just gave us?

OR ANYTHING ELSE.

The Bible story – how close is it to what really happened? Is it fable or history?

A BIT OF BOTH. IN ANCIENT TIMES MY STORY WAS VERY FAMOUS BECAUSE AS A HEBREW IT WAS STRANGE THAT I WOULD HAVE SO MUCH POWER IN EGYPT. BUT IT IS ALSO TRUE THAT MY BROTHERS WERE JEALOUS OF ME AND I WAS SOLD INTO SLAVERY. BUT MY WITS SAVED ME. I WAS NOT THE GREAT SEER SOME THOUGHT I WAS, AND I DID FORGIVE MY BROTHERS THEIR CRIME. THOUGH DETAILS ARE LACKING IN THE BIBLE STORY, IT IS ESSENTIALLY CORRECT. HOWEVER, YOU ARE RIGHT: THE STORY IS TOLD AS A MORAL FABLE. HAD THESE THINGS NOT HAPPENED THE WAY THEY DID – IF FOR INSTANCE I HAD HAD MY BROTHERS MURDERED – THE STORY MIGHT NOT HAVE BECOME PART OF THESE TEACHINGS.

NOW DO NOT THINK THAT JUST BECAUSE MY STORY IS PRETTY ACCURATE THAT ALL THE BIBLE IS SUCH. IT IS NOT. THERE ARE FAIRY TALES MIXED WITH HISTORY AND THEY WILL NEVER BE SEPARATED.

RATHER LIKE ALL OF US HERE. WE ARE ONE. THAT IS THE IMPORTANT LESSON FOR TODAY. JOSEPH SAYS GOODBYE.

Session 37

NERO HERE TO INTRODUCE YOUR GUEST FROM LAST TIME. ALL THAT CONFUSION AT THE END MADE THE WHOLE THING RATHER RAGGED. NOW THAT YOU KNOW WHO YOUR SPEAKER IS, LET HIM SPEAK FOR HIMSELF.

JOSEPH SPEAKS TO YOU NOW. WHAT WAS THE REASON FOR MY BEING CALLED HERE? SOMEONE ELSE COULD HAVE GIVEN YOU MY LITTLE SPEECH ABOUT ONE VOICE. YOU ALSO REALIZED THAT I DID HAVE A FUNNY CONNECTION WITH YOUR COLORS, (the owner of the "coat of many colors") BUT EVEN THAT IS HARDLY ENOUGH REASON FOR MY ATTENDANCE HERE. WHY THEN? BECAUSE MICHAEL WANTED A BIBLICAL FIGURE TO ATTEND THESE MEETINGS. SO MANY CHARACTERS IN THE BIBLE ARE MYTHIC THAT WE ONLY HAD A FEW RELATIVELY INTERESTING ONES TO CHOOSE FROM. SO MUCH OF THE OLD TESTAMENT IS STORY RATHER THAN HISTORY.

SO I WAS CHOSEN, FOR I DID INDEED LIVE. MY LIFE IN EGYPT GAVE ME MORE THAN WEALTH AND POWER. IT ALSO MADE ME AWARE AS ONLY THE OUTSIDER CAN BE AWARE. IT IS WHEN YOU DO NOT BELONG THAT YOU ARE BUSY LOOKING OUT FOR WHAT THE OTHER MAN DOES.

I DO NOT MEAN THAT ONE SHOULD BE SUSPICIOUS, BUT SHOULD BE AWAKE TO WHAT HAPPENS. NO KNOWLEDGE OF THE LANGUAGE IS A WAY TO TURN UPON YOURSELF. ONLY YOU HAVE THE UNDERSTANDING OF YOU IN A VERY FUNDAMENTAL SENSE. YOU PROBABLY NEVER THOUGHT OF MY STORY IN THIS WAY. MOST PEOPLE ARE FASCINATED BY THE CRUELTY OF THE BROTHERS AND MY ABILITY TO MAKE THE MOST OF A BAD SITUATION.

HOWEVER, THAT IS NOT WHAT THIS STORY IS ABOUT. THE THEME, IF THERE IS ONE, IS THAT OF NOT ALLOWING CIRCUMSTANCES TO GET IN THE WAY OF KNOWLEDGE. WISDOM CAME WITH NEW INSIGHTS FROM NEW SURROUNDINGS. NEW IS ALWAYS AN IMPETUS TO RESPONSE, BUT TOO MANY ALLOW FEAR TO GET IN THE WAY OF SINCERE RESPONSE AND THE WORLD MERELY PASSES BY.

AGAIN, THIS IS NOT THE USUAL THEME, BUT ONE OF MY THEMES IS THE NEW, AND THIS IS ANOTHER VIEWPOINT. THIS IS ALSO ONE OF THE

REASONS THE BIBLE WORKS SO WELL: YOU CAN DIG AND DIG IN THESE STORIES AND PULL OUT SOMETHING NEW EACH TIME. THAT IS WHAT MAKES GREAT MYTH.

NOW I WOULD LIKE TO SPEAK OF SOMETHING ELSE: A MEAGER WORD UPON THIS IDEA OF COLOR. LET ME SAY WHAT I SHOULD ABOUT THAT. MY COAT WAS FAMOUS, BUT THAT TOO IS MYTH. WHAT IT IS IS THE VESTIGE OF MY OWN RESPONSE TO COLOR.

IN A SOCIETY WHICH SHUNNED THE IMAGE, WE HAD LITTLE USE FOR PAINT OR PAINTINGS, BUT I SOMEHOW FEEL NOW THAT I ENDED UP IN EGYPT TO FULFILL MY OWN DEEP LONGING FOR ART, WHICH ONLY THEY COULD GIVE ME. THEY HAD AN UNPARALLELED VISUAL SENSE AND WHEN I WENT THERE ALL I WANTED TO DO WAS STAND IN FRONT OF THEIR BEAUTIFUL PAINTED FRIEZES AND DREAM. I KOHLED MY EYES AND WORE THE MOST VIVID BLUE. WHEN MY FAMILY FINALLY CAME TO EGYPT AND FOUND ME, THEY WERE HORRIFIED MORE AT MY HEDONISM THAN AT MY EGYPTIANISM.

THIS RESPONSE, WHICH WAS SO UNLIKE MY OWN PEOPLE, PERSISTED IN THIS COAT. BUT THAT TOO WAS A MYTH. MY YEARS IN EGYPT WERE SUCH THAT I WAS ABLE TO LEARN SKILLS OF PAINTING AND JEWELRY-MAKING AND WEAVING WHICH WERE NOT THOUGHT REMISS IN THE WELL-EDUCATED EGYPTIAN GENTLEMAN. I ALSO SANG AND WROTE VERSE AND THIS LIFE WAS MOST ARTISTIC AND CREATIVE, SOMETHING THAT I COULD NOT HAVE DONE AS A SHEPHERD.

WHAT IS THE POINT? JUST THAT CIRCUMSTANCES MAY BE DECEIVING: WHAT SEEMS AT FIRST A HARDSHIP MAY TURN OUT FOR THE BEST. I WAS GRATEFUL TO BOTH THE GOD OF MY FATHERS AND THOSE I PAINTED IN MY ADOPTED COUNTRY. LIFE IS FULL OF SURPRISES, AND ALL I AM SAYING IS MAKE THE MOST OF THEM. WHAT AN OLD IDEA, BUT HOW RARELY IS THIS ADVICE TAKEN. AND I SAY YOU SHOULD TAKE IT AND USE IT.

NOW JOSEPH LEAVES YOU AND RETURNS TO THIS EMPTY FULLNESS I CALL HOME.

Session 38

(We are discussing E.T.A. Hoffman as we sit down. Tom was reading the stories.)

NERO HERE AND I TOO FIND HOFFMAN AMUSING. HE IS ONE I KNOW. GOETHE INTRODUCED US.

PERHAPS THIS IS WHY HE SUDDENLY INTERESTED YOU. WHO'S TO SAY WHERE WE ALL STOP AND YOU BEGIN? HOW DEEP DOES ONE GO INTO ONE'S MIND BEFORE YOU ARE NO LONGER IN YOUR MIND? THESE ARE PERHAPS FUTILE QUESTIONS AS LONG AS PEOPLE PERSIST IN CREATING BOUNDARIES, BUT THESE ARE AS NOTHING. IT MAY BE ONE MORE WAY TO UNDERSTAND JOSEPH'S SPEECH: THAT WE ARE ONE BECAUSE WE ARE ALL. QUESTIONS?

Tom: What was the problem with Joseph's name?

JOSEPH IS MANY ALSO AND THOUGH HE CAME IN THAT CHARACTER BECAUSE YOU DID NOT KNOW HE DID NOT KNOW WHO HE WAS. THIS IS THE MEANING OF THE MIRROR. YOUR MIND REFLECTS US AND WE MUST BE ABLE TO SEE OURSELVES IN THAT REFLECTION BEFORE WE CAN EXIST FOR YOU.

Tom: What about all those we can't even conceive of – like the spirit of the garden?

ALL THESE FIND REFLECTION IN YOUR MIND. IT IS SOMETIMES A MATTER OF THE VOCABULARY BEING MORE READILY AVAILABLE AND THAT HAS TO DO WITH THE SMOOTH OPERATION OF THE MIND.

WE NEED A CLEAN, CLEAR SURFACE ON WHICH WE MAY BE REFLECTED. AND THESE ABSTRACT CHARACTERS ARE IN SOME WAYS EASIER TO PRODUCE THAN HISTORIC CHARACTERS BECAUSE ONE IS MORE FREE TO EXPRESS HIS ESSENCE IN THE MOST MEANINGFUL WAY, RATHER THAN HAVING TO FOCUS ON A PARTICULAR NAME AND A PARTICULAR LIFE WHEN THE ENTITY NO LONGER FEELS THAT SINGLE EXISTENCE.

Tom: Imagination = Infinity? Is that a clear question?

YES, CLEAR AND IN A SENSE YOU MIGHT SAY THAT YOUR VOICE OF THE VORTEX WAS IMAGINATION IF YOU DO NOT LIMIT THAT WORD WITH ITS USUAL CONTEXTS. IMAGINATION IS A GREAT FORCE AND ITS WORKINGS ARE PART OF OUR GREATEST ENERGY FORMS, BUT THE VOICE YOU HEARD MIGHT BEST BE UNDERSTOOD AS AN EXPRESSION OF MONUMENTAL POWER, WHERE MATTER AND MIND COME TOGETHER. THIS IS AS HE SAID THE SEED OF POTENTIALITY AND THIS IS A CENTRAL IDEA: THE POTENTIALITY OF ENERGY TO MAKE POSSIBLE ALL THE POSSIBILITIES.

SOMETHING ELSE?

Tim: Do we leave intellectual thought behind and let this (i.e., doing the board) flow?

YES, THAT IS PART OF THE PROCESS. BECAUSE YOU HAVE SOME PRECONCEPTIONS OF THE SITUATION YOU BRING A CERTAIN FRAME OF MIND WHICH IS CONDUCIVE TO COMMUNICATION, BUT PERHAPS MORE IMPORTANTLY, THIS SIMPLE MECHANISM FOR THOSE WHO ARE ABLE (AND THAT ABILITY CAN DEPEND ON MANY THINGS) IS TRULY A DIRECT LINK TO COMMUNICATING WITH ENERGY FORMS WHICH ARE INTELLIGENT IN NATURE AND WHICH HAVE NO PHYSICAL FORM.

ONE THING MUST BE MADE CLEAR: THOUGH YOUR MINDS ARE NECESSARY AS A THREAD TO THE REAL WORLD (AS YOU CALL IT) WHAT WE SAY DOES NOT DEPEND ON YOUR KNOWLEDGE OTHER THAN TO SUPPLY WORDS AND FORM OF A SORT. MANY PEOPLE COULD DO THIS TO A DEGREE WHO ARE NOT DOING IT AT ALL, BUT THERE ARE MANY MORE WHO COULD NOT EVEN IF THEY WANTED TO.

Tim: Why is this not popular?

FOR ONE THING, AS YOU HAVE LEARNED, THIS IS NOT EASY. ONE MUST BE WILLING TO EXPEND TIME AND ENERGY AND HOW MANY ARE INTERESTED IN MORE THAN WHO WILL I MARRY AND WILL I BE RICH? IT TAKES, EVEN FOR THIS, A KIND OF DEDICATION THAT NOT ALL HAVE, AND AS YOU KNOW, THE PROCESS – FOR WHATEVER VALUE THE CONTENT MAY HAVE – CAN BE BORING AND TIRING.

Tim: Need open minds too?

YES, OPEN AND TO A CERTAIN DEGREE PASSIVE. AS TOM HAS COME TO UNDERSTAND, THIS IS A QUALITY HE HAS AND HE CAN SUBLIMATE HIMSELF TO THE FLOW OF OUR WORDS. THIS IS THE REASON HE IS SO FLUENT.

Tom: Do you have anything else to say, Nero?

NO, BUT I HAVE. THIS IS NOW ANOTHER VOICE. AND MY NAME IS IRIS. I AM THE RAINBOW. I COME AFTER YOUR MUCH-NEEDED RAIN AND I AM HERE ONLY TO SAY THAT MY COLORS LIGHT YOUR WAY AND ALL YOUR DAYS ARE BLESSED BY SUN AND RAIN. FOR I AM THEIR CHILD. NOW I ARC OVER YOUR OWN ARC OF LETTERS AND SAY THIS TO YOU: BE OF GOOD CHEER, ALL THE POWER OF COLOR IS YOURS TO USE. WE ARE PART OF YOUR ESSENTIAL SELVES. LOOK WITHIN. LOOK WITHOUT. FIND US, SEE US, AND KNOW US. I SHALL SPEAK AGAIN AT THE PROPER TIME. IRIS BIDS YOU *ADIEU*.

Session 39

THIS IS NERO BUT I LEAVE IMMEDIATELY TO HAND YOU OVER TO SOMEONE ELSE.

▣

NOW RABELAIS SPEAKS TO YOU. IT HAS BEEN SOME TIME SINCE WE SPOKE, BUT I HAVE NOT BEEN FORGOTTEN. I HAVE COME TO SPEAK AGAIN ABOUT MY PARTICULAR SUBJECT – ART. AND THIS EVENING WITH A PARTICULAR FOCUS: THAT OF RELATING ART TO COLOR TO PERHAPS DRAW A FEW STRINGS TOGETHER.

NOW IT IS QUITE OBVIOUS THAT COLOR IS IMPORTANT IN PAINTING AND EVEN ANY OTHER PLASTIC ART, INCLUDING ARCHITECTURE, BUT THIS IS NOT THE IMPORTANT COLOR. THE IMPORTANT COLOR IS IN YOUR MIND AND EVEN THEN IT IS NOT A COLOR YOU MIGHT IMAGINE BUT AN EXPRESSION OF EMOTION. ANY WORK OF ART CAN CREATE THIS KIND OF EMOTION BUT IT DOES NOT TAKE AN OBJECT NECESSARILY TO DO IT. WHAT IS NEEDED IS THE RESPONSE TO THE OBJECT.

THIS RESPONSE CAN BEST BE UNDERSTOOD AS A CREATION OF THE MIND OF THE RESPONDER, AND SINCE I SPEAK OF ART TONIGHT I WILL LIMIT MYSELF TO SPEAKING OF THE COLORS OF AESTHETIC RESPONSE. WE AGAIN HAVE THE SPECTRUM, BUT IT IS NOT SO OBVIOUS BECAUSE ONE'S EMOTIONS CAN BE NUANCE AS WELL AS A PUNCH IN THE STOMACH. AND FOR A HAZY RESPONSE THE COLOR IS NOT PRECISE. LET US HAVE AN EXAMPLE.

A PAINTING BY TITIAN FOR EXAMPLE. NANCY AND TOM IMMEDIATELY PICTURED A VAGUE IDEA OF A REMEMBERED TITIAN, BUT TIM RESPONDED WITH THE NAME BECAUSE A PAINTING DOES NOT READILY COME TO HIS MIND.

HOWEVER, BECAUSE THE WORD "PAINTING" WAS MENTIONED THERE WAS THE MENTAL RESPONSE, WHICH HAS NO VISUAL EQUIVALENT EVEN WITHIN ONE'S OWN MIND BUT WHICH CAUSED THE RESPONSE IN TIM'S MIND TO FLUSH FIRST A KIND OF ROSE COLOR WHICH BECAME MORE

GOLDEN, THEN YELLOW, AND FADED INTO WHITE. NANCY, ON THE OTHER HAND, FIRST SAW A DEEP MIDNIGHT BLUE WHICH WAS SUPERIMPOSED OVER A DARK MAROON.

THESE COLORS BLENDED TO A RICH PURPLE WHICH WAS AS MUCH A PRODUCT OF HER RELATING TITIAN TO VENICE AS IT WAS TO THE IDEA OF PAINTING. NOW, I DO NOT MEAN THAT YOU ACTUALLY SAW THOSE COLORS IN YOUR MIND'S EYE. WHAT I OBSERVE IS THE ACTUAL MENTAL PROCESS AND HOW THAT ENERGY IS RELATED TO COLOR. TOM, ON THE OTHER HAND HAD A MOVEMENT ACROSS THE SPECTRUM WHICH BEGAN IN YELLOW AND ENDED IN GREEN. MUCH OF THIS MOVEMENT WAS RELATED TO THE CONCEPT OF ART HISTORY RATHER THAN TITIAN IN PARTICULAR.

EACH OF THESE RESPONSES IS INDIVIDUAL, AND THE POINT OF ALL THIS IS THAT THIS KIND OF FLOWING SPECTRUM IS ONE WAY TO DESCRIBE THE PROCESSES OF THOUGHT AND MEMORY AND CREATIVITY AND ANY OTHER MENTAL PROCESS YOU MIGHT WANT TO NAME.

LET YOUR MINDS CONSIDER THESE THINGS AND WE SHALL SPEAK AGAIN SOON.

Session 40

(Shortly before this session, Tim was offered a job out of the blue which necessitated a move to Chicago.)

NERO HERE AND WE HAVE MUCH TO DISCUSS. ARE YOU READY TO HEAR ABOUT YOUR MOVE? WELL, AS IT HAS SEEMED TO YOU INEVITABLE, SO IT IS. YOU MUST GO AND TOM MUST STAY. THAT IS WHY WE WANTED YOU TO BE CONSCIENTIOUS ABOUT SPEAKING TO US, FOR TIME FOR YOU WAS AT A PREMIUM. WE STILL WOULD LIKE TO HAVE SOME WORK FROM YOU TO CLOSE OFF THIS CHAPTER, BUT NANCY AND TOM WILL HAVE TO GO IT ALONE. WE WILL FINISH BEFORE YOU GO. NOW A BIT ABOUT THE MOVE ITSELF. YOU HAVE BEEN HERE LONG ENOUGH AND BOTH OF YOU NEED STIMULATION, BUT ONE OF THE IMPORTANT THINGS IS THAT THERE ARE PEOPLE YOU MUST COME TO KNOW AND THEY WILL ONLY COME INTO YOUR LIVES IF YOU ARE TRANSFERRED TO CHICAGO.

YOUR IDEA FROM YOUR TIME BEFORE, TIM, WAS NOTHING MORE THAN THE SEED OF THIS MOVE. THE TIME WAS NOT RIGHT; AFTER ALL, NANCY HAD TO BE FOUND FIRST AND NOW YOU CAN RETURN. ANOTHER ASPECT OF THIS MOVE IS MONEY AND THE REAL REASON FOR THAT IS ONE YOU HAVE ALREADY HALF-GUESSED AND THAT HAS TO DO WITH A CHILD WHO WILL BE BORN, WE DO NOT NOW SAY WHEN. BUT START GETTING USED TO THE IDEA. AFTER YOU MOVE WE CAN DISCUSS IT WITH YOU MORE.

DETAILS WILL BE FORTHCOMING, BUT FOR NOW I AM GIVEN LEAVE TO SAY THAT THIS CHILD WILL BE SPECIAL AND YOUR TRAINING HERE IS IN PART DUE TO THIS SPIRIT WHO WILL SOON BE YOURS TO GUIDE.

NOW, DO YOU HAVE ANY QUESTIONS ABOUT ANY OF THIS?

Will this pay off financially over time?

LET THAT TAKE CARE OF ITSELF. AT FIRST YOU WILL CERTAINLY GIVE MUCH ENERGY TO IT, BUT IT WILL LEVEL OFF AND THE JOB I MIGHT SAY HOLDS MANY SURPRISES. YOU ARE SUITED TO THIS SORT OF WORK AND NO YOU ARE NOT TOO MATERIALISTIC. WE ARE. YOU MUST CREATE A SUITABLE ENVIRONMENT AND PART OF THAT IS FINANCIAL SECURITY.

Thanks for the ideas.

YOU WILL DO WELL AND PROSPER. JUST RELAX AND LET IT HAPPEN. YOU ARE MUCH LESS PRONE NOW TO INSECURITIES WHICH GET IN THE WAY OF ACTIONS. YOUR MIND IS MORE AT REST AND THIS WILL COLOR ALL YOU DO, WITHOUT YOUR HAVING TO BE AWARE OF IT.

NOW THEN, LET US TAKE UP THE OTHER SUBJECT ON YOUR MINDS. MR. KOCH, AS YOU CALL HIM, ASKS THAT I SPEAK FOR HIM. (A very elderly friend of Tom's who had just died.) HE DOES NOT HAVE THE MEANS TO SPEAK TO YOU YET BUT HE HAS BEEN TOLD ABOUT THESE COMMUNICATIONS.

What did he think?

ASK NO QUESTIONS. LET ME SPEAK.

MR. KOCH WANTS ME TO SAY THAT HE IS HAPPY NOW AND HE WANTS TO SHARE WHAT HE NOW IS WITH SOMEONE ELSE. BY THIS HE MEANS THAT HE IS BECOMING PART OF A NEW PERSON AND AFTER A BRIEF STAGE HERE HE WILL RETURN FOR ONE LAST LIFE. HE HAD THE SORT THIS TIME WHICH ALLOWS ONE TO HAVE ENOUGH TIME TO GRADUALLY BECOME TIRED OF LIVING. THIS STAGE CREATES A NEW VIEW WHERE THE PERSON CAN RETURN TO LIFE TO ACCOMPLISH GREAT THINGS. MR. KOCH IS AN OLD SOUL AND A VERY FINE ONE AND WE WILL USE HIM TO THE UTMOST. DO NOT WORRY FOR HIM. ALL IS AS IT SHOULD BE. MRS. KOCH HAS A FEW YEARS LEFT AND THOSE YEARS WILL SEE GREAT CHANGES, BUT I LEAVE THAT TOPIC ALONE.

AND NOW ONE LAST THING. THIS BOOK IS FAR FROM COMPLETE, BUT IF YOU AND NANCY COME ON A REGULAR BASIS AS BEFORE, WE CAN COMPLETE THIS SECTION. SOMEONE ELSE LOOMS ON THE HORIZON TO HELP THIS FURTHER ALONG, BUT WE NEED NOT DISCUSS THAT NOW. I HAVE SAID ALL I WANT TO.

NERO SAYS GOODNIGHT.

Session 41

MICHAEL COMES TO YOU THIS EVENING BEFORE THE SUN SETS AND I SPEAK OF THIS WORK. TOM, YOU WILL AGAIN BE WITHOUT PARTNERS IN THIS WORK. NANCY AND TIM, WHO HAVE BEEN SO KIND TO LET US IN, HAVE NEW LIVES AWAITING THEM ELSEWHERE.

THIS WILL STILL BE ONLY A FRAGMENT. WE WILL ROUND IT OFF FOR NOW, BUT MORE IS TO COME. TOMORROW RETURN TO THIS BOARD AND I WILL SPEAK IN THE VOICE OF CASSIEL THE BLUE ANGEL AND I WILL INTRODUCE YOU TO MY BLUE CHILDREN. MICHAEL BIDS YOU *ADIEU*. I GO WITH THE SUN.

Session 43

CASSIEL SPEAKS TO YOU AS THE VOICE OF BLUE. MY REALM IS THAT OF CONTEMPLATION AND WATER AND SKY AND ALL THOSE THINGS WHICH MAKE ONE SAD. BUT LET ME SHOW YOU BLUE FROM ANOTHER ANGLE. BLUE IS THE FORCE WHICH BINDS. IT IS THE TRUE COLOR OF LIFE AND LIKE THE OTHER COLORS A SOURCE OF POWER. IN THIS CASE IT IS THAT POWER WHICH CREATES STARS FROM MATTER AND IN SO DOING MAKES LIFE POSSIBLE. FOR STARS ARE LIKE FACTORIES FOR LIFE. THEIR RADIOACTIVITY IS THE SOURCE OF NEW ATOMS AND THE POWER OF BLUE JOINS THESE ATOMS TOGETHER INTO MOLECULES. AND SO I LET YOU MEET MY FIRST CHILD...

THE MOLECULE...

WHEN ATOMS JOIN IN SPECIAL PATTERNS, THE RESULT IS A MOLECULE. ANY STUDENT OF SIMPLE CHEMISTRY UNDERSTANDS THIS. BUT WHAT CHEMISTRY DOES NOT KNOW IS THAT THESE PATTERNS HAVE MORE MEANING THAN MERE SYMMETRY, THOUGH THAT TOO IS IMPORTANT. THE PURPOSE OF THESE PATTERNS HAS NOTHING TO DO WITH MACRO-LIFE – THAT IS, LARGE CONSTRUCTIONS OF MOLECULES LIKE YOURSELVES – BUT THE REASON IS TWO-FOLD.

FIRST, THE PATTERN IS THAT FORM WHICH ENERGY WILL TAKE TO PRODUCE A PARTICULAR PRODUCT. THOSE SAME ATOMS IN A DIFFERENT PATTERN CREATE DIFFERENT LINES OF ENERGY AND THUS A DIFFERENT MOLECULE. THE OTHER REASON IS FOR MENTAL ENERGY TO BE ABLE TO DISTINGUISH WHAT MOLECULES ARE PRESENT AT ANY GIVEN TIME. THIS MAY NOT SEEM LIKE MUCH, BUT I MUST STRESS THE IMPORTANCE OF PATTERN ON THE ELEMENTS WHICH MAKE COMBINATIONS POSSIBLE, AND THOSE LINES OF FORCE ARE THE POWER OF BLUE. IT IS THE GLUE OF THE UNIVERSE. AND NOW I GO TO LEAVE YOU TO BIG BROTHER, THE GOD OF BIOLOGY...

THE CELL...

THE CELL. SO WHAT? OF COURSE, BIOLOGY CALLS ME ITS SMALLEST UNIT. ALL LIFE IS MADE OF CELLS BUT THEY FORGET THE MYSTERY OF THIS FACT. WHAT AM I? PROTOPLASM, BUT AGAIN JUST A WORD WHICH SIGNIFIES

NOTHING TO MOST PEOPLE. WHAT THEN IS IT IMPORTANT FOR ME TO SAY TO YOU? WELL, IN THE LIGHT OF COLOR, BLUE IS THAT FORCE WHICH HOLDS THE CELL TOGETHER, TOO. NATURE IS SIMPLE IN ITS WAY. WHY HAVE TWO ADHESIVES WHEN ONE WILL DO? BLUE IS THAT ENERGY WHICH KEEPS THE PARTS OF GODS IN RELATION TO EACH OTHER. BLUE IS THE OPPOSITE OF ENTROPY. LIFE IS AN EXCEPTION TO THE SECOND LAW OF THERMODYNAMICS. AND BLUE IS THAT FORCE WHICH MAKES IT POSSIBLE.

BUT AGAIN, WHAT SCIENCE CANNOT KNOW IS THE MYSTERY: THE MIND MADE MANIFEST BY THE CELL. THE MIRACLE OF ASSOCIATION WHICH COMES ABOUT NO MORE AND NO LESS THAN BECAUSE EACH CELL IS COATED WITH ITS DESTINY AND THAT DESTINY IS FULFILLED. DO THE DEAD HAVE CELLS? THE ANSWER SURPRISINGLY IS YES, BUT CELLS OF A DIFFERENT MATERIAL – A "GHOST IMAGE" OF THAT WHICH YOU KNOW. AT DEATH THESE CELLS DO NOT DIE. THEY ARE THE FORM OF THE SURVIVING PERSONALITY. AND WHAT KEEPS IT TOGETHER? BLUE, OF COURSE, THAT SAME GLUE.

AND NOW I SPEAK. I AM THE LAST CHILD OF BLUE.

(Circles)

NOW MY VOICE IS CLEAR. I AM THE POWER WHICH COMES FROM RADIATION. AND WHY AM I A CHILD OF BLUE? I WHO CAN DESTROY ORGANIC LIFE? IT IS BECAUSE I AM ALSO THE SEED OF NEW LIFE. RADIATION I SAY, BUT WHAT DO I MEAN? PARTICLES FROM RADIOACTIVE DECAY. GAMMA RAYS AND SUCH ARE THE FORCE I REPRESENT, AND THOUGH AN ELEMENT OF CHAOS,I AM NECESSARY FOR THE CREATION OF MATTER. YOU MIGHT EVEN SAY THAT I AM THE MOST ELEMENTARY KIND OF CELL: THE CELL OF DESTRUCTION, BUT THE CELL OF NEW LIFE.

CASSIEL RETURNS TO SAY THAT NOW YOU HAVE SEEN MY CHILDREN. AND SO WE MAY PROCEED. BUT THAT COMES WHEN NEXT YOU VISIT THIS BOARD. FOR NOW, MY FAMILY SAYS GOODBYE.

Session 44

TONIGHT WE WILL COMPLETE OUR LECTURE ON REINCARNATION. THIS IS PAN, AND I HAVE MUCH TO SAY. SO LET US BEGIN. YOU HAVE HAD SOME HISTORY OF THE PROCESS AND SOME PERSONAL GLIMPSES AT ITS WORKINGS, BUT WHAT I WANT TO TELL YOU ABOUT IS ITS REAL MOTIVATION.

REINCARNATION URGES MANKIND ON TO SOMETHING HE CANNOT NOW PERCEIVE. THE BASIC IDEA OF EVOLUTION IS CORRECT AS FAR AS IT GOES, BUT SOMETHING MORE IMPORTANT THAN THE RISE AND FALL OF RACES OF CREATURES IS THE ONGOING TRANSFORMATION OF THE EARTH AND THIS IS THE MOVEMENT OF GOD'S MIND.

THE PLACE OF MANKIND IN THIS MIND MUST BE UNDERSTOOD AS A PROCESS OF ENERGY. TOO MUCH TIME HAS BEEN SPENT TRYING TO ANALYZE EACH INDIVIDUAL LIFE. THAT IS NOT SO IMPORTANT.

WHAT IS IS THE ENTIRE MATTER OF ENERGY, WHICH CAN BEST BE UNDERSTOOD AS THE MIND OF MANKIND IN ITS TOTALITY.

THIS MIND IS ONLY ONE OF MANY AND NOT THE MOST IMPORTANT. OTHER MENTAL FORMS ARE HIGHER ON THE EVOLUTIONARY SCALE. SOME OF THESE TYPES YOU HAVE SPOKEN TO AT THIS BOARD.

NOW I AM NOT SAYING THAT EACH INDIVIDUAL IS NOT IMPORTANT, BUT YOU MUST REALIZE THAT EACH IS BUT A FREQUENCY OF ENERGY TUNED OR UNTUNED TO THE HIGHER HARMONIES OF THE UNIVERSE.

BUT ALL IN ALL, IT IS THIS WHOLE STOCK OF HUMAN MIND WHICH IS THE MOST IMPORTANT THING AND WHAT IS INTERESTING IS THAT IT IS OUR JOB TO HELP YOU CREATE SOMETHING ELSE OF THAT MIND.

ONE DAY ALL WILL BE IMBUED WITH THE LIGHT OF PERFECT HARMONY, BUT NOW YOU ARE LITTLE MORE THAN A DISCORDANT TUNE IN A MUCH GRANDER WORK. VERY SOON YOU WILL BE VISITED BY SOMEONE MUCH HIGHER THAN I WHO WILL TELL YOU THINGS THAT I CANNOT.

MY FORM IS THAT OF AN ANCIENT NATURE SPIRIT – A MORE LIMITED VIEW THAN YOUR NEW VISITOR: MOTHER NATURE.

⧈

WELL, WHO ELSE AFTER DEAR PAN? HE IS ONE OF MY OLDEST CREATIONS, BUT ENOUGH NOSTALGIA. THE PURPOSE OF MY VISIT IS TO CLARIFY A FEW THINGS. COLOR HAS BEEN TOUCHED UPON, BUT THE TOPIC IS FAR FROM EXHAUSTED. CONSCIOUSNESS AND THE MIND ITSELF HAVE ALSO BEEN TOUCHED BUT LITTLE MORE THAN THAT. THERE HAVE BEEN STRANGE AND WONDERFUL CREATURES IN YOUR GARDEN THIS YEAR, BUT WHAT DO THEY ALL MEAN?

WELL, PERHAPS VARIETY IS THE SPICE OF LIFE. PERHAPS IT IS A PAGEANT WHICH PASSES ONE BY. CATCH WHAT YOU CAN. YOU FLUNG YOUR NETS WIDE AND LANDED A FEW EXOTIC FISH. WHAT THEN? SERVE THEM UP FOR SUPPER. THE LESSON IS THE LEARNING, CHILDREN. DO NOT LOOK FOR ENDINGS AND BEGINNINGS.

CHANGE, TRANSFORMATION, METAMORPHOSIS – THESE ARE WORDS OF MAGIC AND WONDER. NATURE IS NOT ONLY BOUNTY AND FAMINE, HOT AND COLD. THESE ARE THE WINDS OF CHANGE. TAKE THEN THESE FRIENDS AND THOSE FEW STILL TO COME NOT AS SOME PART OF A FRAGMENTED WORK TO BE FINISHED SOMETIME IN THE FUTURE. EACH CHARACTER IS A CHAPTER, EACH WITH ITS LESSONS AND ENTERTAINMENT. TAKE THEM THEN FOR THIS. USE IT ALL. WHEN YOU RETURN, THE PERSONS YOU WILL BE WILL HAVE CHANGED. SO THEN WILL YOUR VISITORS.

MOTHER NATURE WHO IS ALSO YOUR BLUE QUEEN OF SO LONG AGO BROUGHT CLOSER FOR MORE THAN POETIC AND ZOOLOGICAL INTEREST. [A sly reference to a poem Tom had written 12 or 13 years before.]

THIS HAS BEEN FROM ME TO YOU. MOTHER NATURE WATCHES OVER YOU. I AM IN THE STARS AND IN YOUR HEARTS. BE WELL, CHILDREN.

Session 45

(Renaissance music on the stereo.)

YES, THE MUSIC SUITS ME JUST FINE. MY NAME IS CHRISTOPHER MARLOWE AND WHO BETTER THAN I WHEN YOU REWRITE THE FAUST LEGEND LETTER BY LETTER? NOT MY METHOD. I WAS MORE IMPATIENT THAN THAT. FAUST I SAY, IN WHAT WAY? WAS NOT THAT LEGEND AT LEAST PARTIALLY CREATED TO KEEP THE SEEKERS OF KNOWLEDGE AT BAY? WAS IT NOT SUCH A TREE THAT CAUSED MANKIND'S DOWNFALL? THERE WILL ALWAYS BE THE FEARFUL WHO TRY TO HOLD THE MIND'S RESEARCHES BACK, BUT THERE WILL ALSO ALWAYS BE THE FEARLESS WHO SEEK THE HOLY GRAIL. IS NOT THAT BUT THE OTHER SIDE OF FAUST'S COIN?

MARLOWE, I SAY. TOO MANY NAMES TO RECALL BY NOW, BUT THAT LIFE,THOUGH SHORT,WAS SWEET AND I DO NOT MEAN EASY. SOME LIVES REVERBERATE MORE THAN OTHERS. THE RIPPLES IN ONE'S ENERGY FABRIC ARE MORE FAR-REACHING, AND SEEK THE FURTHER SHORE. SO THIS LIFE I SEE NOW BEFORE ME AS IN A FAULTY MIRROR. WHAT WONDERS OF PERCEPTION AND DECEPTION THEY WORK HERE. MARLOWE THE PEACH-BLOOM POET IS NOT I, BUT I KNOW THAT BOY AS MYSELF.

MARLOWE. WHAT CAN I TELL YOU? SOMETHING SECRET, PERHAPS? LET ME TRY.

ONE OF YOUR SUBJECTS HAS BEEN COLOR. CAN I SHARE SOMETHING ABOUT THAT? I SAID BEFORE THAT THE RIPPLES RUN OUT FROM THAT CENTER OF SELF AND THE REVERBERATIONS CAUSE PERMANENT CHANGE IN THAT SELF. WHAT I DID NOT SAY WAS THAT THESE RIPPLES ARE COLOR, COLOR WHICH IN A SENSE EXPRESSES KNOWLEDGE, OR BETTER, INFORMATION – COLOR-CODED IN THE REAL SENSE OF THAT TERM.

NOW THIS DOES NOT MEAN THAT EACH LESSON HAS A PARTICULAR COLOR. THAT IS MUCH TOO SIMPLE. IT IS RATHER MORE LIKE THE FACT THAT A MIND COMPREHENDS A COLOR INSTANTANEOUSLY AND IN THIS APPREHENSION IS THE COLOR "SEEN."

WHAT I MEAN IS THAT ALL THE INFORMATION IS RECEIVED INTO THE

CONSCIOUSNESS AS A LIVING BEING WOULD SEE RED AND KNOW IT TO BE SO. SO THEN IS INFORMATION GIVEN TO US LIKE A RAINBOW. KNOWLEDGE IS THE SPECTRUM. COLOR IS INFORMATION. FAUST IS AN ARTIST TRYING TO HARMONIZE THE COLORS OF HIS SOUL. SO THEN ARE WE ALL. MANKIND HAS SPOKEN AS BIDDEN.

LET ME LEAVE YOU WITH ONE LAST THING: RED ORANGE YELLOW GREEN BLUE INDIGO VIOLET. THIS IS ALL YOU KNOW ON EARTH AND ALL YOU NEED TO KNOW. THE SPECTRUM IS THE ALPHA AND OMEGA. LEARN YOUR COLORS. BE AN ARTIST OF THE SOUL. HARMONIZE AND MUSIC YOU BECOME.

MARLOWE SAYS GOODNIGHT AND BLESS YOU BOTH.

Session 46

NOW MICHAEL SPEAKS TO YOU AS YOU COME TO THIS PARTING OF THE WAYS. FOR THE TWO OF YOU AND YOU FROM US. IN BOTH CASES THESE WILL NOT BE PERMANENT, BUT CHANGE MUST COME AND SO I WISH TO LEAVE YOU WITH THESE FEW WORDS. MICHAEL IS ONE NAME BUT IN THAT MIND ARE MANY CHARACTERS. YOU HAVE COME TO KNOW SOME OF THEM BUT THERE ARE OTHERS. IN ALL THESE VOICES MICHAEL WAS PRESENT AND THE POINT TO BE MADE IS THAT VARIETY HAS ITS OWN HARMONY AND ALL THESE ASPECTS ARE BUT PART OF A LARGER REALITY OF WHICH I TOO AM BUT A PART.

YOU HAVE BEEN TOLD THAT THIS IS MERELY A FRAGMENT, BUT DO NOT DENY ITS GIFTS EVEN SO. THERE IS MUCH TO BE GLEANED FROM ITS PAGES AND THERE IS ALWAYS MORE ONE CAN LEARN. TOM, YOU HAVE BEEN FEELING APOLOGETIC ABOUT THE LACK OF QUESTIONS YOU HAVE POSED DURING THESE VISITS, BUT DO NOT WORRY. THE TIME WILL COME FOR QUESTIONS. PERHAPS IT IS BETTER THAT YOU MERELY TAKE US IN.

NOW I WILL SAY ONE MORE THING. ON YOUR LAST VISIT YOU WILL SPEAK TO A VOID WHICH FILLS THE UNIVERSE WITH MEANING. LET THIS QUANDARY STAND AS INTRODUCTION ENOUGH. WHEN DO YOU WISH TO MEET?

Tuesday?

TUESDAY WILL BE FINE. LET US SAY SIX P.M. THOSE AT THIS LEVEL NEED A BIT OF CEREMONY. COME WITH YOUR MOST SERIOUS MINDS FOR THIS IS GREAT POWER. NOW MICHAEL SAYS GOODNIGHT AND BLESS YOU BOTH UNTIL NEXT WE SPEAK. ADIEU.

Session 47

NOW I SPEAK TO YOU OF MYSTERIES OF WHICH ONLY I CAN SPEAK. THE STARS AND PLANETS HAVE THEIR OWN LANGUAGE AND THIS LANGUAGE IS COLOR. EACH PLANET IN THIS SYSTEM RELATES TO ONE SPECIFIC COLOR. I GIVE THEM TO YOU NOW.

MERCURY IS RED. VENUS IS ORANGE. EARTH IS YELLOW. MARS IS GREEN. JUPITER IS BLUE. SATURN IS DEEP BLUE AND URANUS IS VIOLET. NEPTUNE IS BLACK AND PLUTO IS TRANSPARENT. THE SUN IS WHITE AS ARE ALL STARS, REGARDLESS OF THE COLOR THEY REFLECT. THIS IS, OF COURSE, THE CASE WITH THE PLANETS AS WELL. THESE COLORS CORRESPOND TO THE SORT OF LIGHT WHICH IS MOST USED BY THE PLANET.

IN EARTH'S CASE, YELLOW IS THE LIGHT OF ORGANIC GROWTH. THESE COLORS ALSO RELATE TO INFLUENCE AND EVOLUTION. JUPITER, FOR INSTANCE, IS BLUE. THIS MEANS THAT THE BLUE RAYS ARE THE NOURISHMENT FOR THAT PLANET AND IT MUST BE ABLE TO UTILIZE MORE YELLOW LIGHT FOR LIFE TO BEGIN. BETWEEN YOU AND JUPITER IS MARS, WHICH IS IN A LIMBO BETWEEN LIFE AND NON-LIFE. ONLY AN INCREASE IN THE UTILIZATION OF YELLOW WILL ALLOW THE PLANET TO PRODUCE LIVING THINGS. STARS ARE WHITE LIGHT, THEREFORE ALL COLORS, AND AS THE SOURCE OF NEUTRINOS WHICH ARE THEMSELVES ALL COLORS, THE FACT IS CLEAR THAT POWER IS COLOR. AND THROUGH COLOR ALL SECRETS ARE POSSIBLE.

THIS SCHEMA IS NOT METAPHORICAL. THESE ARE FACTS BUT FEW PEOPLE ARE INTERESTED IN SEEKING THE TRUE NATURE OF THE UNIVERSE. THE ONLY WAY TO LEARN IS TO SEEK OUT OF YOUR IGNORANCE AND LIKE THE COLORS ON A BRIGHT DAY THAT IGNORANCE WILL BRING YOU KNOWLEDGE.

TAKE THESE WORDS WITH YOU: THINK ON ALL YOU HAVE BEEN TOLD. AND WHEN YOU ARE READY TO SEEK AGAIN WE SHALL BE WAITING. ADIEU MY FRIENDS.

Session 48

NERO HERE TO BID YOU A FOND FAREWELL. I WAS NOT GIVEN A BIG PART IN THE FINAL DAYS OF YOUR PROSE PIECE SO I HAVE COME BACK NOW TO HAVE MY SAY. DID YOU HAVE ANY QUESTIONS, TIM?

How did we get so lucky with the weather – with no snow on the day we loaded our stuff into the truck?

THE WEATHER IS GIVEN TO YOU AS A BLESSING. UNTIL YOU ARE MOVED IN IT WILL CO-OPERATE. IT DOESN'T HURT TO KNOW MOTHER NATURE PERSONALLY. SHE CAN DO PLENTY IN THESE MATTERS. WELL, IF YOU HAVE NO QUESTIONS, I HAVE SOMETHING TO SAY.

THIS BOARD IS A SPECIAL PLACE. NEVER FORGET THAT. IN YOUR DAY-TO-DAY LIVES WE CANNOT SPEAK TO YOU QUITE SO CLOSELY, BUT THAT DOES NOT MEAN THAT WE ARE NOT THERE.

ONE OF THE THINGS THAT THIS WORK HAS DONE IS TO MAKE YOU PART OF ALL OUR MINDS AND SO YOUR ENERGY FIELDS BECOME PART OF OUR OWN FOR ALL TIME. THIS IS SPECIAL TOO BECAUSE IT IS A PLACE FOR COMMUNICATION.

SO MUCH COULD BE SAID AND THERE IS SO LITTLE TIME TO YOU ALL. WE UNDERSTAND THAT YOU HAVE LIVES TO LEAD AND WE CAN ONLY BE A PERIPHERAL EXERCISE, BUT DO NOT LET US REMAIN TOO LONG AWAY FROM YOU. I DO NOT BEG, BUT I KNOW THAT OUR LESSONS TAKE ROOT IN YOUR MINDS. AND AS I SAID, THERE IS MUCH TO BE TOLD.

THE ART OF THIS IS NOT IN THE WORDS PER SE, BUT IN WHAT YOU DO WITH THOSE WORDS. YOUR PAGES NEED NOT SEE A PRINTER TO BE IMPORTANT; EVEN WITHOUT PUBLICATION, IF THEY HELP YOUR LIVES WE HAVE DONE OUR JOBS.

ANY TRANSFORMATION OF ENERGY INTO SOMETHING POSITIVE AND WITHOUT WASTE IS OUR WORK AND ANY LEAST BIT IS LIKE MANNA FROM HEAVEN TO US.

WE ARE MINDS AT WORK AS YOU ALSO ARE. DO NOT FORGET THIS. YOUR BODIES ARE THERE SO YOU HAVE SOMETHING TO STIMULATE YOUR MINDS WITH. USE THEM WISELY AND PERHAPS EVEN TOO WELL.

▣

NOW I SPEAK TO YOU FROM WITHIN THE NUMBER.

NUMBER I SAY. THE FORMULA TO ALL THINGS. PYTHAGORAS IS MY OWN CHILD.

TAKE THIS SMALL GIFT...

(Circles the numbers on the board repeatedly, then circles with more and more power the entire board.)

THESE ARE THE MAGIC SYMBOLS OF ETERNITY. WITHIN EACH IS A MESSAGE FOR UNDERSTANDING THE WORKINGS OF THE UNIVERSE. TAKE THE NUMBER 4, FOR EXAMPLE. THIS IS THE NUMBER OF COMPLETION AND CLOSES A CYCLE. ALL NATURALLY-OCCURRING FOURS HAVE THIS TENDENCY. WHEN A STABLE COMPOUND IS NEEDED, ITS SIMPLEST NUMBER IS FOUR.

SYMMETRY IS DERIVED FROM 4, ALSO. YOU MIGHT THINK TWO WOULD BE ENOUGH, BUT THAT IS A FALSE SYMMETRY. AT LEAST FOUR PARTS IN BALANCED RELATIONSHIP MUST BE PRESENT TO HAVE REAL SYMMETRY.

▣

NERO BACK. THAT WAS A SAMPLE OF SOMETHING ELSE WE COULD TAKE UP AT A LATER DATE. JUST WHEN YOU THINK YOU KNOW IT ALL, THEY THROW IN SOMETHING NEW. WELL, KIDS, THAT'S LIFE.

AND ON THAT NOTE, WHICH IN MY CLEVER WAY RELATES TO YOUR NEW LIVES, GOOD LUCK DEAR FRIENDS. NERO WILL BE HERE WHENEVER YOU CHOOSE TO CALL.

MUCH LOVE. *AU REVOIR.*